Ion
Chromatography

MODERN ANALYTICAL CHEMISTRY

Series Editor: David Hercules
University of Pittsburgh

ANALYTICAL ATOMIC SPECTROSCOPY
William G. Schrenk

APPLIED ATOMIC SPECTROSCOPY
Volumes 1 and 2
Edited by E. L. Grove

CHEMICAL DERIVATIZATION IN ANALYTICAL CHEMISTRY
Edited by R. W. Frei and J. F. Lawrence
Volume 1: Chromatography
Volume 2: Separation and Continuous Flow Techniques

COMPUTER-ENHANCED ANALYTICAL SPECTROSCOPY
Volume 1: Edited by Henk L. C. Meuzelaar and Thomas L. Isenhour
Volume 2: Edited by Henk L. C. Meuzelaar

ION CHROMATOGRAPHY
Hamish Small

ION-SELECTIVE ELECTRODES IN ANALYTICAL CHEMISTRY
Volumes 1 and 2
Edited by Henry Freiser

LIQUID CHROMATOGRAPHY/MASS SPECTROMETRY
Techniques and Applications
Alfred L. Yergey, Charles G. Edmonds, Ivor A. S. Lewis, and Marvin L. Vestal

MODERN FLUORESCENCE SPECTROSCOPY
Volumes 1–4
Edited by E. L. Wehry

PHOTOELECTRON AND AUGER SPECTROSCOPY
Thomas A. Carlson

PRINCIPLES OF CHEMICAL SENSORS
Jiří Janata

TRANSFORM TECHNIQUES IN CHEMISTRY
Edited by Peter R. Griffiths

Ion
Chromatography

Hamish Small

Formerly Research Scientist
The Dow Chemical Company
Midland, Michigan

Plenum Press • New York and London

Library of Congress Cataloging in Publication Data

Small, Hamish.
  Ion chromatography / Hamish Small.
     p.   cm — (Modern analytical chemistry)
  Includes bibliographical references.
  ISBN 0-306-43290-0
  1. Ion exchange chromatography. I. Title. II. Series.
QD79.C453S63 1990   89-39790
543'.0893 — dc20   CIP

© 1989 Plenum Press, New York
A Division of Plenum Publishing Corporation
233 Spring Street, New York, N.Y. 10013

Printed in the United States of America

To Beryl, Deborah, and Claire

Preface

Bewitched is an odd word with which to begin a chemical textbook. Yet that is a fair description of how I reacted on first learning of ion exchange and imagining what might be done with it. That initial fascination has not left me these many years later, and it has provided much of the motivation for writing this book. The perceived need for a text on the fundamentals of ion chromatography provided the rest.

Many readers will have a general idea of what ion chromatography is and what it does. Briefly, for those who do not, it is an umbrella term for a variety of chromatographic methods for the rapid and sensitive analysis of mixtures of ionic species. It has become highly developed in the last decade, and while it is now routinely used for the determination of organic as well as inorganic ions, its initial impact was greatest in the area of inorganic analysis. In the past the determination of inorganic ions, particularly anions, meant laborious, time-consuming, and often not very sensitive "wet chemical" methods. In the last ten years that has changed radically as ion chromatography has supplanted these older methods.

While ion chromatography (IC) is relatively new, the principles that underlie its practice and guide the development of its several parts have been laid down over decades of research in many diverse areas of chemistry. A primary purpose of this book is to provide a systematic account of this fundamental chemistry and the principles that underlie it.

Ion chromatography is a practical art, and this book is about practical matters. So although there is an emphasis on basic principles, I have placed equal importance on showing the direct relationship of the underlying scientific concepts to everyday practical matters of IC, such as the design and synthesis of ion exchange resins and the solution of real chromatographic problems. I have treated detection in IC at some length, in the first place because detection stands equal in importance to separation in modern chromatography, and second because new methods of detection are largely responsible for the renaissance of ion exchange in analytical chemistry. The book has also given me the opportunity to address the specific area of conductometric detection, which, in IC, is still a source of some misunderstanding.

To whom is the book addressed and to whom might it appeal? Modern instrumentation, with its emphasis on operator convenience and its ability to automate, has been a tremendous boon to the analyst in increased productivity and decreased tedium. On the other hand, it tends increasingly to insulate the user from the underlying science, so texts and the like that emphasize the theoretical bases of methods provide a necessary counterbalance to this "black box" tendency. I believe such texts to be especially necessary in chromatography, where the operator is such an active participant in the analytical process, in that decisions must often be made on such matters as separation method, stationary and mobile phases, and modes of detection. So in considering to whom the book should appeal I always had in mind the practicing chromatographer. I hope also that the book will be of help to those who are involved in the design of new chromatographic phases since I have endeavored to treat the synthesis of exchangers and the origins of ion selectivity at some length. And an early chapter on the chromatographic process is intended as a bridge that practitioners of LC and GC will find useful in making the transition to ion exchange chromatography.

In the research phase of *Ion Chromatography,* and particularly in the writing of the book, I have been keenly aware of the rich hinterland from which we drew much of our resources. I hope the book goes at least a short way in recognizing those who, though they could not know it at the time, have contributed so significantly to the birth and success of this most recent offspring of Ion Exchange.

ACKNOWLEDGMENTS. I am especially indebted to Dr. Ed. Johnson and Ted Miller for their critical reading of the entire manuscript and for the many improvements that they suggested. Also my sincere thanks go to Rosanne Slingsby and Nancy Jensen for their help with many of the figures.

Hamish Small

Leland, Michigan

Contents

Chapter 4

Ion Exchange in Ion Chromatography

Chapter 5

Ion Exchange Resins in Liquid Partition Chromatography

Chapter 9

Selected Applications of Ion Chromatography

APPENDIXES

Chapter 1

Introduction

The analysis of complex mixtures often requires that they first be separated into their components. There are a variety of reasons for this: similarity in the chemistry of the components is one. Even where analytes are few, they can be so overlapping in properties that probes or methods of adequate selectivity are lacking and separation is required. In these cases separation is an essential part of the analytical procedure. But even when the mixture contains species so distinctly different in chemistry that selective methods can be devised for their individual measurement, there can be other compelling reasons for making separation a part of the total analytical procedure—speed of analysis, for example. Should one or more of the various determinations be time consuming, then the total time to analyze a single sample can be considerable. A preliminary separation, on the other hand, if it is fast enough, opens up the possibility of applying nonselective measurement techniques to the separated components. Then, if the measurement step matches the speed of the separation, the coupling of the two can provide an effective solution to the time problem. Modern day chromatography exemplifies this successful marriage of separation and essentially instantaneous measurement of the separated components. This book is devoted to a particular part of chromatography, the chromatographic separation and measurement of ionic species, or ion chromatography (IC) as it is better known.

1.1. BACKGROUND TO ION CHROMATOGRAPHY

The chromatography of ions as practiced today is a result of the merging of two major areas of development, chromatography and ion exchange. Deciding when the two came together is a subjective judgment at best, but there can be little doubt that the work of Adams and Holmes[1] was an important landmark. They were the first to demonstrate the possibility of synthetic ion exchangers by forming cross-linked, insoluble polymers from phenols, phenylene diamines, and formaldehyde. These polymers had the ability to exchange ions and were

1

vastly superior in chemical stability to the natural zeolites that were the most widely known ion exchange materials of that time. Major improvements on this pioneering work of Adams and Holmes quickly followed. I.G. Farbenindustrie patented condensation polymers of phenol sulfonic acid and formaldehyde that were useful over a much wider range of pH than were the original phenol formaldehyde resins of Adams and Holmes. And in 1945 a patent granted to D'Alelio[2] that described the sulfonation of polystyrene pointed the way to chemical modification of neutral cross-linked polystyrene resin as being a powerfully versatile route to any number of ion exchange resins and ultimately to chelating resins. These events were the foundations of a new industry directed at the synthesis and application of ion exchangers, endeavors in which many companies both in Europe and in North America were to participate.

The early 1940s was a lean period for publications on ion exchange, but while these war years may have halted the flow of information, their effect on research and development in this exciting new area was quite the opposite. Stimulated by the needs of the Manhattan Project, ion exchange and particularly ion exchange chromatography enjoyed intense and exceedingly fruitful development. Some of the work of that period was eventually declassified and reported in a special section of the *Journal of the American Chemical Society* in 1947. Describing the theory and application of ion exchange resins, this collection of papers is surely a classic of the literature of ion exchange. There we find the first reports by Spedding, Tompkins, and others[3,4] of the chromatographic separation of the rare earth elements. Their use of complexing agents such as citrate to amplify the subtle chemical differences in these closely similar ions was the forerunner of much of present day practice in the chromatography of metal ions. Bauman and Eichhorn[5] and Kunin and Myers[6] introduced mass action concepts to ion exchange, described the ion exchange equilibria of the two major types of ion exchange resin, and determined equilibrium constants for several ion exchange reactions. Boyd, Adamson, and Myers[7] realized the diffusional nature of the ion exchange process and their theoretical and experimental approaches have been the foundation for much of the work on ion exchange kinetics in the years since. Mayer and Tompkins[8] were among the first to tackle the complex theoretical challenge of describing the ion exchange column separation process. Marinsky, Glendenin, and Coryell among others[9,10] were the first to apply ion exchange chromatography to the separation of radioactive species and to couple it directly to radiometric detectors.

The years immediately following World War II were a golden age for ion exchange. For about two decades, researchers worldwide contributed prodigiously to the understanding and applications of these fascinating new artifacts, ion exchange resins. Thousands of articles and many books recorded the myriad developments.[11-15] This period was also marked by another significant event in the development of ion exchange chromatography, the separation and automated detection of amino acids, an achievement by Moore and Stein[16,17] that would eventually earn them the Nobel prize in chemistry in 1972. This, and subsequent work by Hamilton and co-workers,[18-20] presaged much of modern liquid chromatography in that it was an early demonstration of the synergistic coupling of a chromatographic separation with a continuous flow-through detector.

Besides these significant practical developments there was also progress on the theoretical side of ion exchange chromatography. Glueckauf's work on relating column performance to the fundamental properties of ion exchange resins can be singled out as a major contribution that anticipated many of the theoretical concepts of modern high performance liquid chromatography.[21]

Parallel to these advances in ion exchange, chromatography was having profound effects on analytical chemistry, particularly of organic materials. The development of gas chromatography provided a rapid, selective, and sensitive means of analyzing complex mixtures of closely similar, volatile organic compounds. Later, important advances in the use of liquid mobile phases extended chromatography to the whole range of organic materials whether they were volatile or not.

Besides the advances in separation, the other key to chromatography's success was the concurrent development of automated detectors. In fact, by the early 1970s, chromatography had in many instances become synonymous, not with separation alone, but with separation coupled to simultaneous detection. And since a great many organic compounds absorbed in the UV part of the spectrum, one of the principal detectors that provided this prompt determination of separated species was the flow-through UV photometer. It enabled the detection of species—including ionic species—as diverse as proteins, dyes, pharmaceuticals, and synthetic polymers and became the workhorse detector of high-performance liquid chromatography.

Although the UV detector was the most widely used, it was not the only detector. If compounds lacked light-absorbing properties, then postseparation reactions could often be devised to yield chromophoric products that absorbed in the visible, or, failing that, properties such as fluorescence or electrochemical activity often afforded a means of quantifying the separated analytes. But many organic materials lacked any of these properties. Refractive index is a universal property, but detectors based on it lacked the requisite sensitivity. So for these various reasons many organic species were deemed unamenable to chromatographic analysis.

What of the chromatography of inorganic species? The analytical chemistry of inorganic species is in large part the analysis of electrolytes in aqueous solution and to this ion exchange brought considerable benefits. Many ion exchange chromatographic schemes were devised that provided clean separations of inorganic ions, both cationic and anionic. But although the separations were effective, they were used mostly as an adjunct to existing wet chemical methods in order to solve interference and matrix problems. As a result, the speed of the analysis was often determined by the relatively slow wet chemical method. When photometric and electrochemical detectors became available, there were some attempts to apply them to the chromatographic analysis of some inorganic ions—transition metals and lanthanides, for example.[22,23] In general, however, by the early 1970s, chromatography as applied to inorganic analysis was, comparatively speaking, moribund.

Although there may have been a number of reasons for this stagnation, a major one was technical in nature—the lack of a good universal detector for inorganic ions. Many common ions did not absorb visible or UV radiation to any useful extent: alkali metal and alkaline earth metal ions, and anions such as fluoride, chloride, sulfate, and phosphate, for example. Others such as bromide, nitrate, nitrite, and iodide were light absorbing, but in wavelength regions that were inaccessible to the earliest UV detectors. Thus, the normal photometric methods of UV/visible detection were deemed unsuited to the measurement of many inorganic ions. Nor could convenient postseparation chemistries be devised that would generate chromophores and other properties such as fluorescence and electrochemical activity were generally lacking in this very important class of ions.

In late 1971, at the Dow Chemical Company, research was started that eventually led to the adoption of conductometric detection as a universal means of quantifying inorganic ions and indeed of a great many organic ions as well. Ion exchange chromatography was used to separate the ions but the total procedure included the novel addition of a second column after the separator that modified the effluent prior to passing it to a conductivity cell. This second column, the "stripper," later called the "suppressor," essentially removed the background eluent but allowed the analytes to remain, often with enhanced conductivity. This solved the problem of detecting small amounts of electrically conducting analytes against a background of highly conducting mobile phase. The inventors of the "eluent suppression" approach made the first report of their research in 1975.[24]

Earlier in that same year, the Dow Chemical Company, which had applied for patents on the new technique, granted a license to the Dionex Corporation to manufacture and market instruments that embodied the suppressed conductometric approach. They called the technique "ion chromatography," and the first "ion chromatograph" instrument was displayed at the 1975 fall meeting of the American Chemical Society in Chicago. In the years immediately following, Dionex pioneered the commercialization of ion chromatography and, in collaboration with their customers, established the market for IC, often in areas that could not be foreseen by the original inventors. This combination of invention and marketing changed the face of inorganic ion analysis and much of organic ion analysis as well, for now instruments were available that could determine a wide diversity of ions with a speed and sensitivity that had been unattainable by the older classical methods of analysis.

As the market for instruments for ion chromatography grew, so also did alternative methods. In 1979, conductometric methods that did not use a suppressor were first reported[25,26] and commercialization of the nonsuppressed approach began, first by the Wescan Company[27] and later by Waters, Shimadzu, Metrohm, and others. At about the same time there was an evolutionary change in the meaning of the term ''ion chromatography.'' Previously it had been applied solely to the eluent suppression technique, but with the increasing prominence of other techniques for the chromatography of ions, the name, logically, and happily, came to embrace a much wider range of methods. Nowadays the term ''ion chromatography'' includes any chromatographic method applied to the determination of ions. It is even applied retrospectively to include older methods such as the chromatography of amino acids.

The research on conductometric detection brought other benefits besides the direct one of easier detection. The new ion exchange resins that were devised to make suppressed conductometric detection feasible, brought ''fall-out'' benefits in that they eliminated the slow elution and sluggish mass transfer that were characteristic criticisms of the older, high-capacity materials. These improved properties stimulated their use in conjunction with other detectors besides the conductometric varieties.

Conductometric detection was coupled with other methods of separation besides ion exchange. Paired ion approaches were linked to both suppressed and nonsuppressed methods. Ion exclusion, originally developed for large-scale separation, was adopted by analytical chromatographers and now enjoys a prominent place in the total repertoire of IC methods.

Today the chromatographic analysis of ionic materials is widely applied and rapidly expanding. The number of species that may be determined continues to grow, as does the number of areas of science and technology where IC plays an important role. Table 1.1 gives some idea of the breadth of application of ion chromatography at the present time.

TABLE 1.1. Types of Samples Analyzed by Ion Chromatography

Acid rain	Ores
Analgesics	Pesticides
Chemicals	Pharmaceuticals
Detergents	Physiological fluids
Drinking water	Plating baths
Fermentation broths	Protein hydrolysates
Fertilizers	Pulping liquors
Foods and beverages	Soil and plant extracts
High-purity water	Wastewater

1.2. THE BOOK

1.2.1. Aims

The primary aim of this book is to identify and define the basic principles that guide the practice of ion chromatography. To this end the book describes the fundamental bases of the major separation and detection techniques of IC; discusses the problems presented by the disparate requirements of separation and detection; shows how these problems have been solved in a compatible and harmonious way; and brings out the relationship between theory and practice through many examples from a wide range of IC applications.

1.2.2. Organization

Chapter 2 deals with chromatographic separation. It describes the process of chromatographic separation in a general way, avoiding emphasis on any particular mode of separation. It deals exclusively with column chromatography and explains the basis of band displacement. The reader is encouraged to consider the selective partitioning of solutes between two *static* phases as a preliminary to the more complicated case of partitioning between phases in relative motion. This is particularly helpful in understanding the rate of band movement when several chemical forms of a species are involved, each individually governed by a unique set of dissociation and partitioning equilibria. This is often the case in ion exchange chromatography.

Band broadening and its fundamental causes are discussed, and the chapter concludes with a brief discussion of resolution in chromatography.

Chapter 3 describes the materials that comprise the stationary phases most commonly used in IC. Here the emphasis is on ion exchange materials, particularly ion exchange resins and the special forms that had to be developed to make IC practical. The chapter presents methods of synthesis of the main generic types of ion exchange resins and some of their most important physical and chemical properties. There is also a brief discussion of ion interaction reagents.

Chapter 4 is in large part a description of the ion exchange process and its relationship to chromatography. It treats such fundamental features as ion exchange selectivity and equilibria, the Donnan potential, complexation and ion exchange equilibria, and the rate of ion exchange and its relationship to band broadening. Chromatographic topics treated include the various types of ion exchange chromatography, elution and displacement development and frontal analysis, peak shape in elution development, and the causes of various base-line disturbances.

Although detection is not treated in any detail in this chapter, this topic is broached from time to time in the context of how ion exchange behavior influences and dictates the method of detection.

Chapter 4 concludes with a section on ion interaction reagents in IC. Ion interaction systems display behavior that has much in common with ion exchange, so this was a logical place to discuss them.

Chapter 5 treats a number of IC methods that use ion exchange resins or closely related materials but not in an ion exchange mode. Ion exclusion, crown-ether-based resins, and a variety of other novel stationary phases are described.

Chapters 6, 7, and 8 deal with detection in IC. Chapter 6 is a general treatment of the problems of detection in chromatography. It emphasizes the importance of noise in measurement and how it may be reduced. A considerable part of the discussion is devoted to sensitivity of detection in chromatography and to its proper definition.

Chapter 7 deals exclusively with conductometric detection in IC. The subject is treated in a way that shows the evolution of conductometric detection and the rationale behind the two principal methods, suppressed and nonsuppressed. The chapter concludes with a comparison of the two approaches, which is augmented with a computer simulation of their application to an ion analysis problem.

Chapter 8 describes the other major methods of detection that have been applied in IC, including so-called indirect methods.

Chapter 9 provides several examples from the literature of IC applications that illustrate the separation and detection principles discussed throughout the book.

1.2.3. Terminology

The book uses many terms that, with perhaps one exception, are commonly used in the fields of ion exchange and chromatography. For the benefit of those that are new to either or both of these fields, each term is at least briefly defined.

Eluite is a term that may not be familiar to the reader. It is used to denote any sample species that is eluted or in the process of elution from a chromatographic column. Borman attributes the term to Horvath,[28] who coined the word as a more precise alternative to either "solute" or "analyte." As he pointed out, describing the species injected in a sample as solutes has ambiguities since anything dissolved in anything else is by definition a solute. Analyte is a better term for describing the species in a sample, especially if the objective of the chromatographic experiment is an analytical one. But what of a species that is injected for some other purpose—to establish its retention time for instance? Clearly the objective in that case is physicochemical rather than analytical, and the "neutrality" of purpose implied in the term eluite makes it a more appropriate choice than the alternatives. For such reasons I favor the term eluite and have used it liberally throughout the book.

REFERENCES

1. B. A. Adams and E. L. Holmes, Adsorptive Properties of Synthetic Resins. I, *J. Soc. Chem. Ind. (London)* **54**, 1-6T (1935).
2. G. F. D'Alelio, Ion Exchangers, U.S. Patent No. 2,366,007 (1944).
3. F. H. Spedding, A. F. Voight, E. M. Gladrow, and N. R. Sleight, The Separation of Rare Earths by Ion Exchange I. Cerium and Yttrium, *J. Am. Chem. Soc.* **69**, 2777–2781 (1947).
4. E. R. Tompkins and S. W. Mayer, Ion Exchange as a Separation Method. III. Equilibrium Studies of the Reactions of Rare Earth Complexes with Synthetic Ion Exchange Resins, *J. Am. Chem. Soc.* **69**, 2859–2866 (1947).
5. W. C. Bauman and J. Eichhorn, Fundamental Properties of a Synthetic Cation Exchange Resin, *J. Am. Chem. Soc.* **69**, 2830–2836 (1947).
6. R. Kunin and R. J. Myers, The Anion Exchange Equilibria in an Anion Exchange Resin, *J. Am. Chem. Soc.* **69**, 2874–2879 (1947).
7. G. E. Boyd, A. W. Adamson, and L. S. Myers, Jr., The Exchange Adsorption of Ions from Aqueous Solutions of Organic Zeolites. II. Kinetics, *J. Am. Chem. Soc.* **69**, 2836–2849 (1947).
8. S. W. Mayer and E. R. Tompkins, Ion Exchange as a Separations Method. IV. A Theoretical Analysis of the Column Separations Process, *J. Am. Chem. Soc.* **69**, 2866–2874 (1947).
9. J. A. Marinsky, L. E. Glendenin, and C. D. Coryell, The Chemical Identification of Radioisotopes of Neodymium and of Element 61, *J. Am. Chem. Soc.* **69**, 2781–2786 (1947).
10. E. R. Tompkins, J. X. Khym, and W. E. Cohn, Ion Exchange as a Separations Method. I. The Separation of Fission-Product Radioisotopes, Including Individual Rare Earths by Complexing Elution from Amberlite Resin, *J. Am. Chem. Soc.* **69**, 2769–2777 (1947).
11. F. Helfferich, *Ion Exchange,* McGraw-Hill, New York (1962).
12. O. Samuelson, *Ion Exchange Separations in Analytical Chemistry,* Wiley, New York (1963).
13. J. Inczédy, *Analytical Applications of Ion Exchangers,* Pergamon Press, Oxford (1966).
14. W. R. Rieman and H. F. Walton, *Ion Exchange in Analytical Chemistry,* Pergamon Press, Oxford (1970).
15. H. F. Walton, *Reviews on Ion Exchange,* published bienally in *Analytical Chemistry* (1966 to 1980). (An excellent resource for information on all aspects of ion exchange, theoretical and applied.)
16. S. Moore and W. H. Stein, Photometric Ninhydrin Method for Use in the Chromatography of Amino Acids, *J. Biol. Chem.* **176**, 367–388 (1948).
17. S. Moore and W. H. Stein, Chromatography of Amino Acids on Sulfonated Polystyrene Resins, *J. Biol. Chem.* **192**, 663–681 (1951).
18. P. B. Hamilton, Ion-Exchange Chromatography of Amino Acids. Study of Effects of High Pressures and Fast Flow Rates, *Anal. Chem.* **32**, 1779–1781 (1960).
19. P. B. Hamilton, D. C. Bogue, and R. A. Anderson, Ion-Exchange Chromatography of Amino Acids. Analysis of Diffusion (Mass Transfer) Mechanisms, *Anal. Chem.* **32**, 1782–1792 (1960).
20. P. B. Hamilton, Ion-Exchange Chromatography of Amino Acids—A Single Column, High Resolving, Fully Automatic Procedure, *Anal. Chem.* **35**, 2055–2064 (1963).
21. E. Glueckauf, Principles of Operation of Ion-Exchange Columns, *Ion Exchange and its Applications,* pp. 27–38, Society of Chemical Industry, London (1954).
22. Y. Takata and G. Muto, Flow Coulometric Detector for Liquid Chromatography, *Anal. Chem.* **45**, 1864–1868 (1973).
23. J. S. Fritz and J. N. Story, Chromatographic Separation of Metal Ions on Low Capacity Macroreticular Resins, *Anal. Chem.* **46**, 825–829 (1974).
24. H. Small, T. S. Stevens, and W. C. Bauman, Novel Ion Exchange Chromatographic Method Using Conductometric Detection, *Anal. Chem.* **47**, 1801–1809 (1975).
25. K. Harrison and D. Burge, Anion Analysis by HPLC, Abstract No. 301, Pittsburgh Conference on Analytical Chemistry (1979).

26. D. T. Gjerde, J. S. Fritz, and G. Schmuckler, Anion Chromatography with Low-Conductivity Eluents, *J. Chromatogr.* **186,** 509–519 (1979).
27. T. H. Jupille, D. W. Togami, and D. E. Burge, Single-Column Ion Chromatography Aids Rapid Analysis, *Ind. Res. Dev.* **25**(2), 151–156, February (1983).
28. S. Borman, Eluent, Effluent, Eluate, Eluite, *Anal. Chem.* **59,** 99A (1987).

Chapter 2

The Chromatographic Process

2.1. INTRODUCTION

Chromatography today exists in such a variety of forms that defining it in a way that is concise as well as comprehensive is virtually impossible. Such a definition should recognize that what we call the mobile phase in chromatography can be a gas, a liquid, or a supercritical fluid, and that it may contain electrolytes or other modifiers necessary for the separation process. Such a definition should accommodate all the manifold forms of the stationary phase: solids, gels, liquids immobilized in solids, coatings on the walls of capillaries, and even those cases that appear to involve no stationary phase at all. And an adequate definition should convey some idea of the variety of ways in which the two phases are presented to each other: in columns, as a thin layer on a plate, as a paper strip suspended in a reservoir of solvent, etc.

Indeed, the challenge of adequately defining chromatography or even one of its subbranches is so formidable that the result would probably be so exceedingly cumbersome as to be of little use. A better approach is to avoid all encompassing definitions and instead to develop a knowledge and understanding of the unifying concepts that link all the branches of the methodology and science that is chromatography.

To this end this chapter will provide a broad description of the chromatographic process and introduce concepts, terms, and definitions that are common to many areas of chromatography including ion chromatography. Later chapters will discuss the phenomena and behavior that are peculiar to the chromatography of ions.

2.2. A DESCRIPTION OF A CHROMATOGRAPHIC SEPARATION

Ion chromatography belongs to that broad subclassification of chromatography known as liquid chromatography. The term ''liquid chromatography'' (LC), as used in this book, is understood to imply at least two constraints: (1) that a liquid is used as the mobile phase and (2) that the stationary phase is contained in some sort of envelope such as a column or capillary. We will begin with a description of how liquid chromatography might be practiced and introduce some commonly used terms.

A key element of a liquid chromatograph is the column, a tube packed with some sort of solid or gel in a finely divided form, commonly and desirably spherical. This packing is referred to as the **stationary phase.** A porous plug or filter supports the packing and prevents it washing from the tube.

The interstitial or void volume of the column, i.e., the space between the packing particles, is filled with a liquid, the **mobile phase,** that is continuously pumped through the column.

In a typical chromatographic operation a volume of mobile phase containing species to be separated is injected into the flowing mobile phase and carried into the column, at which time separation begins. If the objective is to recover the separated species, then one will provide some means of collecting fractions of the **eluate,** as the exiting fluid is called, and these fractions will subsequently be processed to recover the hopefully purer components of the injected mixture. On the other hand, if the objective is a purely analytical one, that is, to learn something of the composition of the injected mixture, then modern practice will place some sort of detecting device at the column exit to monitor species as they are **eluted** from the column. The response of such a detector might be as depicted in Figure 2.1 if the injected sample contained just two species in addition to the components of the **eluent,** as the mobile phase is termed in this context.

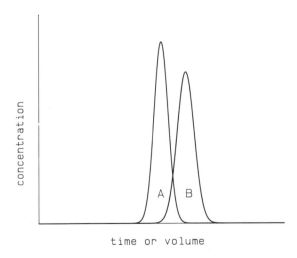

FIG. 2.1. Partial chromatographic separation of two species.

The most significant observation, of course, is that a differential migration has taken place within the column that has resulted in a partial separation of the injected species. Although we will usually strive for a more complete separation than that depicted here, it is the separation, however imperfect, that is the essence of chromatography.

This chapter will address two fundamental questions: Why does the separation take place, and what factors control the shape of the bands? The phenomena that relate most directly to these questions divide conveniently into two distinct areas:

1. The chemistry and thermodynamics associated with the distribution of solutes between two contiguous phases. It is the unequal partitioning of species between phases that is responsible for the centers of mass of the solute bands moving apart as they do.
2. Kinetic and hydrodynamic effects that are the dominant causes of the broadening of the bands.

2.3. THERMODYNAMIC ASPECTS OF CHROMATOGRAPHIC SEPARATION

2.3.1. Distribution of a Solute between Two Phases

The distribution of solutes between phases, as well as being of great intrinsic interest, is one of the core concepts of chromatography, and our success in the practice of chromatography will depend to a great extent on our fundamental knowledge of these distribution equilibria and on our ability to influence them.

Consider two contiguous phases, which, in anticipation of a later context, we will refer to henceforward as the stationary and mobile phases (Figure 2.2). As to the chemical composition of these phases, the only restrictions to be assumed for the moment are (1) that the mobile phase is a liquid and (2) that the stationary phase is some sort of solid or gel.

When a third component, denoted the solute, is added to the system it will distribute between the phases, and, given sufficient time, an equilibrium will be established that may be described by a **partition** or **distribution coefficient,** denoted K_D, such that

$$K_D = C_s/C_m \tag{2.1}$$

where C_s and C_m are the concentrations of the solute in the stationary and mobile phases.

If the distribution of solute was determined on purely statistical grounds then K_D would always be unity or, stated in another way, the fraction of the total solute in either phase would be equal to the volume fraction of the system

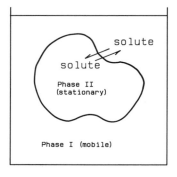

FIG. 2.2. Distribution of solute between two phases in contact.

occupied by that phase. But the distribution is not purely a random event; it is instead biased by the "chemistries" of the two phases, and as a result K_D in general differs from unity. For systems where the solute has a high affinity for the stationary phase K_D can be very large, while values of K_D close to zero will be characteristic of systems where the solute greatly prefers the mobile phase.

In later parts of the book we will be interested in how the distribution of a solute may be manipulated through changes in the chemistry of the two phases, but of immediate interest is the fact that for a particular pair of phases the distribution coefficients of chemically different solutes are not equal. This selectivity in the partitioning process is one of the most basic and important concepts of chromatography: it is in fact a prerequisite for chromatographic separation. We will now examine how this separation comes about.

2.3.2. Chromatographic Separation of Solutes

Another way of treating distribution recognizes the rapid and chaotic motion of molecules of solute between the two phases and considers that the equilibrium reflects the average time that a solute molecule spends in one or the other of the two phases or, alternatively, the probability of finding it in a particular phase at any time. Thus of two solutes, the one with the larger K_D will spend a greater fraction of its time in the stationary phase than will the one with a smaller K_D.

Consider now a device consisting of a column filled with particles of stationary phase over which mobile phase is pumped. If a small volume of mobile phase containing two solutes A and B is injected into the flowing mobile phase then A and B will be carried into the column and begin the process of partitioning between the phases. Now time spent in the stationary phase does not contribute to a species advancement through the column, while time in the mobile phase advances the solute at the rate of the mobile phase. So if species B has a larger K_D than A, then B will lag behind A since the former spends a larger fraction of its time in the stationary phase. In time both solutes will elute from the column

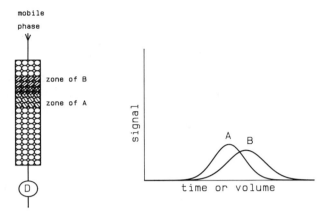

FIG. 2.3. Chromatographic separation of species A and B. B has a higher K_D than A and is therefore retarded relative to A.

but displaced in time in a manner that might be represented by the overlapping peaks (Figure 2.3). [An eluted solute, or one in the process of elution, will often be referred to as an *eluite* (see also Chapter 1 on terminology).]

The relationship between K_D, the elution volume of an eluite, V_E, and the volumes of stationary and mobile phases, V_s and V_m, respectively, is one of fundamental importance to chromatography. We will now derive their relationship in a somewhat intuitive way.

We can ascribe a volumetric flow rate to the eluite as being that volume swept out per unit time by a plane through the center of mass of the eluite band. In order to clear the column, an eluite band will have to sweep through a volume V_m, so the elution time will be

$$t_E = V_m/\text{flow rate of eluite} \qquad (2.2)$$

Now the eluite flow rate will equal the mobile phase flow rate, denoted R, only when the eluite is completely excluded from the stationary phase. If it spends any fraction of its time in the stationary phase then its apparent velocity will be less than R since, as we have already seen, time spent in the stationary phase is time lost to movement through the column. If the distribution of eluite was a purely statistical phenomenon (i.e., no chemical effects) then its apparent flow rate would be

$$RV_m/(V_m + V_s) \qquad (2.3)$$

What of the case when K_D is not equal to 1? K_D reflects the chemical (physical) difference between the two phases with respect to the solute but a mathematically equivalent way of considering the system is to assume that the phases are chemically identical (i.e., $K_D = 1$) but that the volume of the stationary phase has an *apparent* volume V_{app} that is related to its real volume as follows:

$$V_{app} = K_D V_s \qquad (2.4)$$

That is, the volume of the stationary phase appears to be inflated or diminished relative to its actual volume depending on whether K_D is greater or less than unity. Consequently, the apparent flow rate of an eluite of distribution coefficient K_D is

$$RV_m/(V_m + K_D V_s) \qquad (2.5)$$

The time to clear the void volume of the column is then

$$t_E = (V_m + K_D V_S)/R \qquad (2.6)$$

or

$$t_E R = V_m + K_D V_s \qquad (2.7)$$

The quantity $t_E R$ is the elution volume, V_E, of the eluite, so (2.7) becomes

$$V_E = V_m + K_D V_s \qquad (2.8)$$

This expression is one of the most important and fundamental ways of describing the transport of a solute through a chromatographic column. It relates the elution volume directly to a thermodynamic quantity K_D that may in turn be obtained from equilibrium measurements on static batch systems. Rarely, however, in the practice of ion chromatography, or indeed of any form of chromatography, is the chromatographic experiment preceded by a measurement of K_D, so one of the objectives of this book will be to attempt as far as possible to show the reader how K_D in ion chromatography systems may be calculated from some fundamental properties of the stationary and mobile phases, and perhaps more importantly to predict trends in K_D as the properties and composition of the two phases are changed. Only when we have mastered this can we make the most effective use of the chromatograph. Fortunately, the decades of research on the fundamental properties of ion exchange materials have provided modern day ion chromatographers with a sound foundation on which to base these predictions. This is not the case in many other forms of chromatography, where the experimental design often involves a good deal of empiricism.

2.3.3. Relationship between k' and K_D

Another important descriptive term in chromatography is the **capacity factor** or, as it is better known, the **k-prime**. k' is defined as

$$k' = \frac{\text{Moles of A in stationary phase}}{\text{Moles of A in mobile phase}}$$

where A is some partitioning solute. Thus

$$k' = V_s C_s / V_m C_m \qquad (2.9)$$

and since $C_s/C_m = K_D$ then $k' = K_D V_s/V_m$.

Combining equations (2.8) and (2.9) gives

$$k' = (V_E - V_m)/V_m \qquad (2.10)$$

k' is obviously a simple and experimentally more accessible term than K_D; obtaining it involves only the determination of two readily measurable quantities. It is, however, more of an operational definer of the system in that it is a *post factum* description of the chromatographic experiment. Although k' is a less fundamental factor than K_D it is nevertheless a commonly accepted way of describing the behavior of eluites in chromatography and it will be widely employed throughout this book.

2.3.4. Variation of K_D with Concentration—Nonideal Behavior

When K_D of a system comprising two phases and a solute is measured over a wide range of solute concentration, rarely is K_D found to be independent of solute concentration; hence the practice of referring to it as a distribution coefficient rather than a distribution constant. Only in the most ideal systems will K_D be invariant over a wide range of concentrations, and such systems are the exception rather than the rule in chromatography and nowhere more so than in ion exchange systems. Nevertheless, as will be seen presently, there are regions of concentration in which K_D is essentially constant. We will now develop the formal thermodynamic description of this nonideality and examine its implications for chromatography.

An ideal solution is one where the chemical potential of each component is related to its mole fraction or concentration, C, thus

$$\mu = \mu^0 + RT \ln C \qquad (2.11)$$

where μ^0 is the chemical potential of some standard or reference state whose value is dependent only on temperature and pressure. In an equilibrium system such as depicted in Figure 2.2 the thermodynamic definition of equilibrium is that state when the chemical potential of solute in the two phases is equal, i.e.,

$$\mu^0_s + RT \ln C_s = \mu^0_m + RT \ln C_m \qquad (2.12)$$

which may be rearranged to

$$\ln(C_s/C_m) = (\mu^0_m - \mu^0_s)/RT \qquad (2.13)$$

At constant temperature and pressure the term on the right-hand side is a constant and equation (2.13) can be expressed as

$$C_s/C_m = K_D = e^{\text{const}} \qquad (2.14)$$

This is often referred to as the distribution law or partition law of distribution equilibria, and occasionally as Nernst's law.

Stationary and mobile phases in chromatography are usually sufficiently dissimilar that if a solute forms ideal solutions in one phase it is unlikely to form them in the other. This nonideal behavior may be expressed either by the isotherm, which is a function relating the concentrations in the two phases, or by activity coefficients. An activity coefficient, denoted γ, is a measure of how a component deviates from ideal behavior and allows us to use equations formally similar to those that apply to ideal systems. Thus the general equation for chemical potential is

$$\mu = \mu^0 + RT \ln \gamma C \qquad (2.15)$$

when at unit activity coefficient (an ideal solution) it becomes identical to equation (2.11).

We can now write an expression for a true thermodynamic equilibrium distribution constant as

$$K_D^* = \gamma_s C_s / \gamma_m C_m \qquad (2.16)$$

If we had extensive knowledge of activity coefficients over a wide range of concentrations for solutes in both phases then we would have a ready means of calculating K_D's. While this information is occasionally available for liquid phases—electrolyte solutions are an important example of this—it is rarely

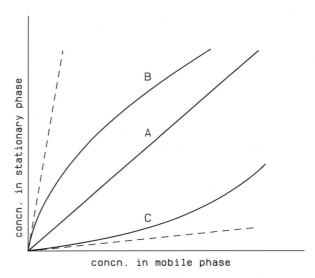

FIG. 2.4. Types of isotherms for the distribution of a species between two phases.

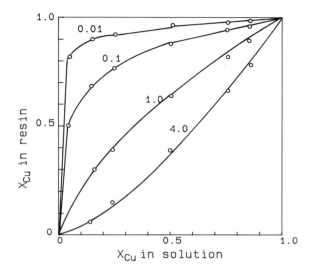

FIG. 2.5. Distribution of cupric ion between ion exchange resin and solution in the system Dowex 50/NaNO$_3$: Cu(NO$_3$)$_2$, at various total molarities. Values on the curves refer to total molarities. (Reproduced by permission of the American Institute of Chemical Engineers from Ref. 1. Dowex is a trademark of the Dow Chemical Co.)

available for stationary phases such as solids or for gel phases such as ion exchange resins. It is more common therefore to assess ideality from isotherms.

In the present context, an isotherm is a function that relates concentrations in a two-phase multicomponent system at constant temperature. The isotherm is commonly expressed in graphical form (Figure 2.4). Retaining the notation of mobile and stationary phases, ideal behavior is illustrated by isotherm A, where through a range of concentration of solute, the concentration of solute in one phase is proportional to its concentration in the other. The slope of the line is K_D and it is constant.

Few systems in chromatography are this ideal; a more likely situation is depicted by isotherm B, where the slope declines with rising concentration, i.e., K_D is concentration dependent and is lower the higher the concentration. Every system will, however, tend towards ideality as shown by the asymptotic approach of isotherm B to the dotted line that represents a Henry's law type of behavior. As will be seen shortly, there is a benefit to operating under chromatographic conditions where Henry's law is applicable, that is, in the linear region of the isotherm.

Type B isotherms are common in chromatographic systems and especially in ion exchange systems; Figure 2.5, for example, shows isotherms for the exchange of copper for hydronium ion on an ion exchanger.

A less common type of isotherm is represented by curve C of Figure 2.4, where K_D diminishes with diminishing concentration; the sorption of weak acids into strong acid form cation exchangers (Chapter 5) is an important example of this type of behavior.

2.3.5 Isotherms and Peak Shape in Chromatography

When a solute is eluted through a chromatographic column the profile of its concentration as a function of distance along the column is often as illustrated in Figure 2.6A. A later section will address the problem of peak shape and broadening, but at this point we will anticipate some of those observations by stating that it is a matter of experience that the peaks are often symmetrical and have the shape of a normal or Gaussian distribution function. For example, solutes that are in an ideal range of behavior, i.e., K_D constant, will tend to give symmetrical peaks. On the other hand, deviations from ideality will lead to two forms of "skewing" of the band. A type B isotherm will lead to a feature known as "tailing" (Figure 2.6B), while type C leads to "fronting" of the peaks (Figure 2.6C). The following is an explanation for these skewing effects.

Even for a solute with a nonlinear isotherm, the peak profile will not be severely distorted when the solute has penetrated but a short distance into the column. As a first step, therefore, assume that at this early stage of development the concentration is approximately symmetrical (Figure 2.7A). In nonideal systems K_D depends on concentration. This means that, in a chromatographic con-

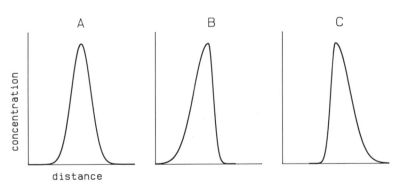

FIG. 2.6. Peak skewing. Profiles of eluite concentration *within* the column as a function of distance from the column inlet. A, Profile for constant K_D. B, Profile from a type B isotherm. C, Profile from a type C isotherm. The shape of the eluted peaks as revealed by the detector will be as follows: type A, symmetrical; type B, sharply rising front, more gently dropping tail (tailing); type C, gently rising front, sharply dropping tail (fronting).

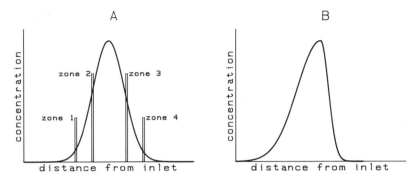

FIG. 2.7. Basis of skewing. Illustrated for a type B isotherm, where K_D of the eluite increases with decreasing concentration. Profiles represent concentrations within the column as a function of distance from the column inlet.

text, a concentration zone will move at a rate that is dependent on the concentration within that zone. The consequences for peak shape for the two types of isotherms now become obvious. For a type B isotherm, since lower concentration zones exhibit a higher K_D, zone 1 (Figure 2.7A) will tend to lag relative to zone 2, which is above it on the concentration profile. Consequently, the band "upstream" of the band center will tend to become more diffuse or "strung out" as elution proceeds. Conversely, on the "downstream" side of the peak median, zones of lower concentration such as zone 4 resist elution relative to zones such as zone 3. This has a self-sharpening effect on the down or exit side of the peak since zones tend to "bunch up." In due course a profile such as Figure 2.7B will develop within the column and a linear detector placed at the outlet of the column will reveal the sharply rising arrival of the solute peak followed by the broad tailing off that is characteristic of solutes that obey a type B isotherm. By a similar argument it can be shown that a type C isotherm will lead to peaks having just the opposite shape, that is, a diffuse leading edge but a sharply falling trailing edge.

Peak broadening, as we will see presently, is an inevitable feature of chromatography; it is also an undesirable one since it impairs the separation of peaks. Peak skewing is therefore particularly undesirable since asymmetrical peaks broaden more rapidly than symmetrical ones and the more convex (or concave) the isotherm the more pronounced and detrimental the effect. In later sections we will learn of the several other factors that influence peak broadening, but at this juncture we can recognize a benefit from operating in a range where K_D is independent of solute concentration. Arranging that this be so is one of the challenges of chromatography.

2.4. DYNAMIC ASPECTS OF THE CHROMATOGRAPHIC PROCESS

2.4.1. Band Broadening in Chromatography

If thermodynamics were the only controlling factor in chromatography then separations could be effected on very short columns indeed. To illustrate, consider the separation of two species whose k' values are, respectively, 1.1 and 1.15, the column volume is 1 ml, the packing is spherical so that V_m is approximately 0.35 ml, and the injected volume containing A and B is 0.01 ml. If the separation of the solute bands took place *without any band broadening* then a detector of A and B placed at the column outlet would give two sharp square wave pulses of A and B (Figure 2.8) each 0.01 ml in width, their centers of mass 0.0175 ml apart as calculated from (2.10). The two solutes differing only 5% in their distribution coefficients are completely separated—in fact the system is overdesigned in that complete disengagement could be accomplished in a somewhat shorter column in this hypothetical system.

Real systems do not behave in this way with all molecules displaying identical elution times; they display instead a certain distribution of elution times about the mean so that the reality of chromatography is more correctly represented by the Gaussian shaped peaks. In this example the distribution coefficients, column volume, etc. are identical to those of the previous hypothetical experiment but "an efficiency of 1000 plates" has been assumed for the column under the prevailing chromatographic conditions. The terminology, efficiency and plates, may be unfamiliar; suffice it to say that it is a quantitative means of expressing the phenomenon of band broadening that is an inevitable part of the chromatographic process.

Now it is evident in this more true-to-life example that the species are very

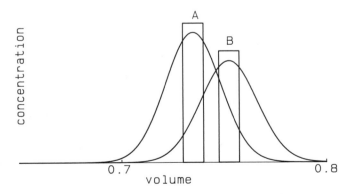

FIG. 2.8. Comparison of "ideal" and real chromatograms. Ideal (no kinetic effects) is represented by the rectangular peaks. Real is represented by the Gaussian peaks. (Note: rectangular peaks and Gaussian peaks are on different scales.)

far from being completely separated, and it fittingly prompts the question: What can be done to improve the separation?

A successful strategy must obviously involve measures that move the centers of mass of the peaks farther apart. From the earlier discussion of equilibrium phenomena we know that this can be accomplished in one of two ways: either by increasing the length of the column, which will give a proportionate increase in the interpeak volume, or by altering the chemistry of the eluent/packing combination so that the difference between the K_D values is increased. There are no *a priori* grounds that guarantee that either measure will be successful since we do not know as yet how the peaks broaden as we take these other steps.

2.4.2. Band Broadening—Some Experimental Observations

Before taking up the theoretical basis of band broadening it is appropriate to summarize some of the more significant experimental findings:

1. For a solute with "linear" elution, that is, K_D independent of concentration, the shape of the peak is approximately that of a normal or Gaussian distribution.

2. The farther a solute is eluted, that is, the longer the column, the greater the peak width. More specifically, the peak width is proportional to the square root of the length of the column. (A constant cross section is assumed in these comparisons.)

3. When the chemical and physical properties of the mobile and stationary phases are fixed and the column parameters are fixed, then the peak width is found to depend in a complex way on the flow rate of the mobile phase, sometimes being independent of flow rate but generally diminishing with decreased flow rate.

4. When the particle size of the stationary phase is changed, all other things remaining the same, the peak width diminishes with decreasing particle size.

It has been recognized for some time that chromatography is a separation technique based on equilibrium phenomena but run under nonequilibrium conditions. Since it takes a finite time for the mobile and stationary phases to equilibrate, and since the mobile phase is applied in a continuous manner to the column, it is evident that the concentration status in a narrow zone within the column will always be displaced from the status appropriate to an equilibrium condition. So if we consider the case of a peak that has already undergone some spreading, the concentration of the mobile phase downstream from the peak maximum will be somewhat greater than the appropriate equilibrium value, while the opposite obtains above the maximum. Putting this in terms of k' or K_D one might say that zones 1 and 2 of Figure 2.9 are moving under the influence of "pseudo"-k-primes that are, respectively, higher and lower than that pertaining to the maximum, which is the k-prime appropriate to the equilibrium condition. Consequently there is a tendency for zones 1 and 2 to pull apart as elution

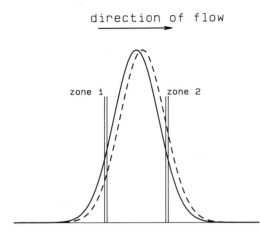

direction of flow

zone 1   zone 2

FIG. 2.9. In a real chromatographic system, equilibrium is disturbed due to flow. The concentration profile of eluite in the mobile phase (dashed curve) leads that of the eluite in the stationary phase (solid curve). In a column run under equilibrium conditions the two profiles would overlap.

progresses, which in turn would lead to broadening of the peak. Since one would intuitively expect then that broadening would depend on the degree of departure from equilibrium, it can be expected to increase with increasing flow rate, and if we assume that the rate of interphase transfer is diffusional in nature then the effect of changing particle size is at least qualitatively explained.

And what of the Gaussian shape of peaks and the dependence of peak width on column length? These and other effects are not so intuitively obvious and we must now turn to the two major theoretical approaches that address these nonequilibrium issues in a quantitative way.

2.4.3. The Plate Theory of Band Broadening

The plate theory is not so much a theory of chromatography as it is an apt analog for the chromatographic process. It was originally developed by Martin and Synge[2] who were the first to recognize the complementarities of behavior between an actual chromatographic column run under nonequilibrium conditions and a hypothetical column that is subdivided into a number of stages that are operated at equilibrium. The following relatively simple treatment of such a multistage device will reveal the main features of this model.

Consider the contacting device depicted in Figure 2.10 consisting of a number of unit cells containing mobile and stationary phases. The device is operated in a series of steps wherein the contents of the mobile phase from cell 0 are moved to cell 1 and thence to cell 2 etc. while the stationary portions of the cells remain fixed.

Now consider a molecule placed in cell 0 and after equilibration the mobile phase of cell 0 is advanced to cell 1. If p represents the probability of the

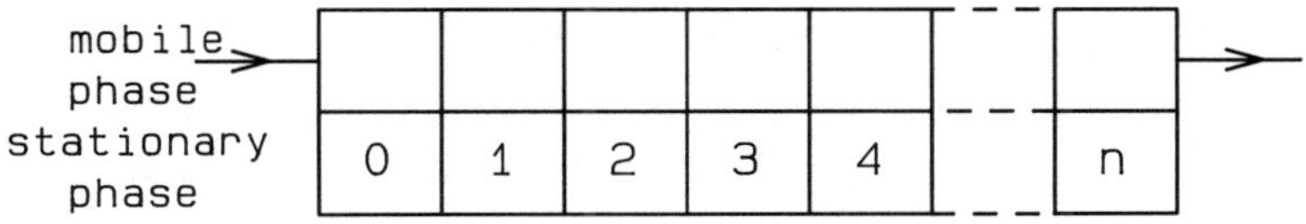

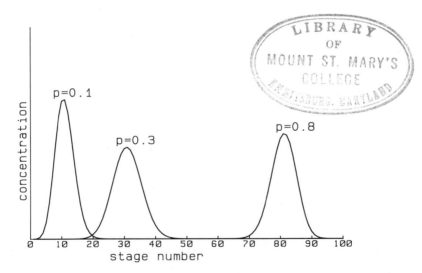

FIG. 2.10. Hypothetical contactor. Each cell comprises an element of mobile phase and one of stationary phase. The contractor operates by allowing equilibration between elements of a cell, then advancing the mobile phase elements one unit to the right, reequilibrating, then advancing the mobile phase elements another unit, and so on. An equilibration followed by one movement to the right is called a transfer.

molecule being distributed into the mobile phase then it also represents the probability that it will be transferred into cell 1. Or, considered in another way, one transfer will advance the molecule through p stages so that the stage number (M) in which the molecule is most likely to be found after r transfers is given simply by $M = rp$.

For a large number of molecules stage M is the stage in which the concentration of molecules would be maximum.

Figure 2.11 shows how the position of this stage of maximum probability varies with the value of p for 100 transfers. We will set aside for the moment the question of the shape of the curves about the maxima and how they are calculated; our objective for the present will be to express the movement of the mean in terms of some familiar chromatographic terms, viz., V_E, V_m, V_s, K_D, and k'.

FIG. 2.11. Distribution of solutes of different p values throughout a contactor of a limitless number of stages, after 100 transfers.

2.4.3a. Relationship between K_D, V_E, V_s, and V_m

Instead of considering the device as a limitless number of stages it is considered to consist of n stages. The number of transfers r required to advance the mean concentration of a species to the nth stage is given as

$$rp = n \qquad (2.17)$$

The probability p can be expressed as the ratio

$$\frac{\text{Amount of species in mobile phase of cell}}{\text{Total amount in cell}} = C_m v_m / (C_m v_m + C_s v_s) \qquad (2.18)$$

where C_m and C_s are, respectively, the equilibrium concentrations in the mobile and stationary phase portions of a unit cell, and v_m and v_s are, respectively, the volumes of the mobile and stationary portions. Dividing the numerator and denominator of (2.18) by C_m gives

$$p = v_m / (v_m + K_D v_s) \qquad (2.19)$$

If V_e denotes the total volume of the stationary phase that has moved in order to bring the peak maximum into the nth stage then it is related to v_m by the expression

$$V_E = r v_m \qquad (2.20)$$

Combining (2.17), (2.19), and (2.20) gives

$$V_E = n v_m / p = n v_m + n K_D v_s \qquad (2.21)$$

Since $n v_m$ and $n v_s$ are the total volumes of the mobile and stationary phases in a device of n stages therefore

$$V_E = V_m + K_D V_s \qquad (2.22)$$

Notice that this is identical to the relationship that was derived by a somewhat intuitive means for a chromatographic column (2.8).

2.4.3b. Dispersion of Solute in a Multistage Contactor

So far we have considered the movement of the maximum or mean through the multistage contactor and have observed that the total volume of mobile phase required to move the maximum through a specified number of stages is proportional to the number of stages and inversely proportional to p (2.17) and (2.21). The dispersion of the band is also dependent on p (Figure 2.12) and on the number of stages through which the solute is driven (Figure 2.13) but to a degree that is obviously different from the movement of the peak maximum. If we were in fact to measure the widths of the peaks in Figure 2.13 we would find that for

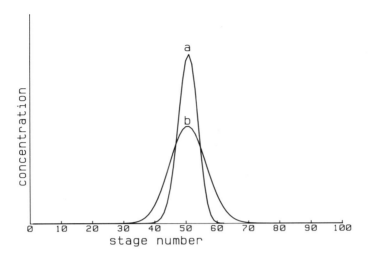

FIG. 2.12. Distribution of two solutes (a) $p = 0.8$ (b) $p = 0.3$, after the number of transfers necessary to bring each maximum to the 50th stage.

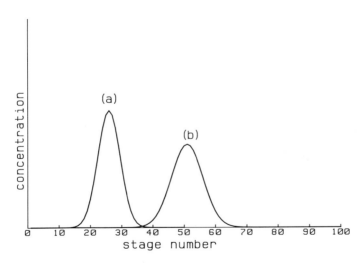

FIG. 2.13. The concentration of a solute ($p = 0.5$) in a contactor after (a) 50 transfers and (b) 100 transfers.

the twofold increase in the movement of the peak position, the width of the peak has only increased by a factor equal to the square root of 2. We will now show how this dispersion is derived and how it closely resembles the concentration profile of a chromatographic peak, and how, in turn, the characteristics of such a hypothetical device may be used to describe the dispersion of solute in a real chromatographic column.

Consider that we place in the first cell of the hypothetical contactor a unit amount of some solute (Figure 2.14). After equilibration the mobile phase portion of cell 1 is advanced into cell 2 and fresh mobile phase is advanced into cell 1. We will define this combination of equilibration followed by advance as a transfer. Now if p represents the fraction of solute remaining in the mobile phase and q the fraction transferred to the stationary phase, then the distribution, cell by cell, after one transfer will be as shown in Figure 2.14. The phases are now reequilibrated with the result that fractions p and q of the total contents of each stage (cell) will distribute into the mobile and stationary phases, respectively, so that the second transfer will give a result as depicted.

As this operation is continued it becomes apparent that the distribution of solute among the stages can be represented by the expansion of the expression $(q + p)^r$, where the total concentration in each stage is given by each term of the expansion. Or, returning to the single molecule experiment, each term of the expansion represents the probability of finding the molecule in a particular cell after r attempts to get it there. Also from earlier arguments we have seen that its most likely locus is the rpth cell.

The binomial expansion of the expression $(q + p)^r = 1$ provides for calculating this probability for any cell, $n;$ thus

$$P_{n,r} = r!q^{r-n}p^n/n!(r-n)! \tag{2.23}$$

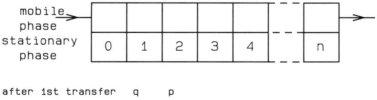

FIG. 2.14. Distribution of solute in each cell of a contactor after first, second, and third transfers. The distribution after r transfers is given by the binomial expansion $(q + p)^r = 1$.

expresses the probability that the molecule will be in the nth cell after r transfers. This general binomial expression provides a means for constructing distribution diagrams as depicted in Figures 2.11–2.13.

2.4.3c. Dispersion as Given by the Binomial Expansion

From probability theory[3] the standard deviation about the mean for such a binomial distribution is

$$\sigma_s = (rpq)^{1/2} \tag{2.24}$$

For a particular solute where the product pq is constant, (2.24) reduces to

$$\sigma_s \propto r^{1/2} \tag{2.25}$$

Now since the length (volume) of the device through which solute is driven is proportional to the number of transfers effected, (2.25) transforms to

$$\sigma_s \propto L^{1/2} \tag{2.26}$$

We have already noted that band broadening in real chromatography is proportional to the square root of column length, so (2.26) illustrates one of the ways in which the plate model is in harmony with the real chromatographic system that it represents.

We have seen that the distance traveled by the peak maximum is proportional to the number of transfers, while the broadening of the band is proportional to the square root of the number of transfers. This has profound implications for the success in separating since it says that the peak maxima are pulling away from each other at a greater rate than the peaks are broadening.

2.4.3d. The Effect of Stage Size (Plate Height) on Band Broadening

Let us assume a device of fixed length or volume and examine the relationship between broadening, σ_s, and the number of stages, N, into which the device is subdivided.

From (2.25) we have $\sigma_s \propto r^{1/2}$. Now r is proportional to N since it takes proportionately more transfers to move the solute maximum to the Nth stage as N is increased. So

$$\sigma_s \propto N^{1/2} \tag{2.27}$$

This states that the dispersion of the peak *in stages* increases with the number of stages, but since the volume per stage is inversely proportional to N, the volumetric dispersion, σ_v, is given by

$$\sigma_v \propto 1/N^{1/2} \tag{2.28}$$

This clearly shows that broadening is diminished as the hypothetical device is subdivided into smaller and smaller elements. The corollary of this for a real chromatography column is that sharper bands are indicative of columns operating as if they had a greater number of theoretical plates.

2.4.3e. The Gaussian Profile

The shape of the bands in Figures 2.11–2.13 resembles one of the familiar functions of probability theory; in fact, as n becomes large the binomial distribution tends to a Gaussian distribution, another way in which the model matches experience.

Thus, in terms that are appropriate to a chromatographic operation, the Gaussian expression linking concentration and volume of the emerging peak is

$$C = [1/\sigma(2\pi)^{1/2}][\exp[-0.5(V - V_E)/\sigma]^2] \qquad (2.29)$$

where V_E is the elution volume of the solute and σ is the peak dispersion. Measurement of the peak dispersion therefore provides a means of assessing the number of "theoretical plates" in a column or alternatively the height equivalent to a theoretical plate (HETP). The relationship for calculating plate number N is

$$N = (V_E/\sigma)^2 \qquad (2.30)$$

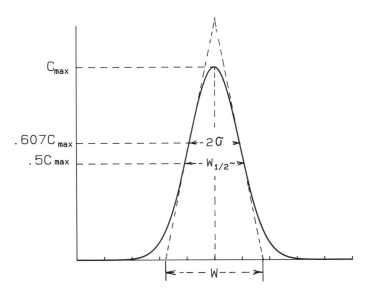

FIG. 2.15. Various methods of determining the standard deviation from a Gaussian shaped peak. $W_{1/2} = 2.354\sigma$; $W = 4\sigma$.

The HETP is simply the column length divided by N. An alternative expression for H is

$$H = \sigma^2/L \tag{2.31}$$

where L is the distance migrated by a peak and σ is the standard deviation of the peak dispersion in units of length. Figure 2.15 gives the common ways of determining σ from a Gaussian distribution. In practical terms, σ is conveniently and probably most accurately assessed from $W_{1/2}$, the width at the half height, since applying tangents at the inflection points can be somewhat subjective.

2.4.3f. Limitations of the Plate Theory

Although the plate model of a chromatographic column gives results that are in excellent harmony with experience, it must be emphasized that it has some serious deficiencies. The most serious criticism is that it has very little true predictive value. While it does reveal correctly the relationship between band broadening and length it is unable to predict the magnitude of the dispersion— the plate height in other words. Nor is it able to predict the effect of changing operating variables such as flow rate or particle size since it takes no account of these important factors. Perhaps its greatest merit is its relative simplicity.

2.5. RATE THEORIES OF BAND BROADENING

Whereas the plate theory shows how a continuous process may be described in terms of a discontinuous one and has some success in doing so, we have also observed the limitations of such an approach. Its major deficiency is its disregard for the phenomenological details of the process that it describes. Rate theories, on the other hand, confront these details and attempt to connect band spreading with such properties of the system as velocity of the mobile phase, particle size of the packing, packing irregularities, and the rate of mass transfer between the phases.[4-7] The rate approach is therefore a more realistic and consequently a more valid approach to the problem of band spreading.

2.5.1. The Origins of Band Broadening

There are three contributors to band spreading: longitudinal or axial diffusion, eddy diffusion, and resistance to mass transfer between phases.

Longitudinal Diffusion. When a zone of solute is introduced into a column it is immediately influenced by the natural process of diffusion and loses its initial sharpness as the solute diffuses symmetrically about its center of mass as it is carried through the column. How serious is this type of broadening? Not very serious it turns out—at least not for packed columns in liquid chromatography. For fast eluting solutes, that is, solutes that spend most of their time in the mobile phase, their time in the column is usually so short that not sufficient time elapses from injection to exit for much diffusional dispersion to take place. We can obtain without great difficulty an upper limit for the spreading by solving the simple problem of the longitudinal dispersion of an infinitely thin disk of solute into contiguous solvent. The solution to the problem[8] states that the concentration as a function of distance from the point of origin and of time is given by the expression

$$C = [M/2(\pi Dt)^{1/2}]\exp(-x^2/4Dt) \tag{2.32}$$

where M is a constant representing the amount of material in the disk, D is the diffusion coefficient, x is the distance from the point of origin, and t is the time.

This is yet another form of the familiar Gaussian function from which we can deduce that the standard deviation is $(2Dt)^{1/2}$. Now if L is the length of the column and U the linear velocity of the mobile phase then $t = L/U$ and

$$\sigma = (2DL/U)^{1/2} \tag{2.33}$$

It turns out that for the mobile phase velocities normally used in LC, this contribution to band spreading is very small. If one also keeps in mind that this diffusional model assumes unrestricted diffusion into the mobile phase, whereas in a column diffusion takes place in the presence of barrier effects from the packing particles, then the estimate provided by (2.33) is conservative in the sense that it exaggerates the spreading.

Of course some solutes will remain in the column for much longer times, but since they are spending most of their time in the stationary phase where diffusion coefficients are usually markedly lower than in the mobile phase, longitudinal dispersion within the packing is correspondingly low.

So axial diffusion is not a significant contributor to band spreading in liquid chromatography and what little there is is inversely dependent on the rate of flow of the mobile phase.

Eddy Diffusion. This is the term applied to dispersion caused by splitting of streams by the column packing. This is a complex hydrodynamic problem, so a simplified model is helpful. Consider the migration of a molecule carried in a solvent through the hypothetical packing depicted in Figure 2.16, where the direction of flow is vertically downwards. We will assume for simplicity that the system is two dimensional. The molecule arriving at point A has two choices of direction in which it may go—to point B or to point C. Either direction will produce the same result as far as advancing the molecule in the direction of flow. Now consider the molecule arriving at point B and confronted with the same choice; obviously a migration towards D will contribute less to its transport in the direction of flow than taking a path towards point E. In a large assembly of

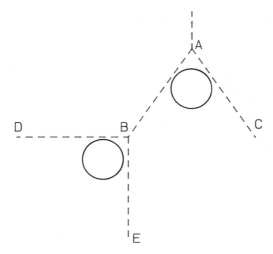

FIG. 2.16. Random pathways of solute through a bed of spherical particles.

molecules confronted with a large number of barriers (choices) there is clearly a degree of randomness to the pathways that they travel through the bed. Some molecules will take more advantageous pathways through the bed and arrive at the outlet ahead of the average while some will take less advantageous routes and arrive after the mean, thus leading to a distribution of arrival times. The spreading that results from this type of random flow is commonly termed eddy diffusion.

What can be deduced from this model about the relationship of eddy diffusion to other properties of the chromatographic system? Intuitively one would expect that the dispersion should be independent of flow rate through the column since the number of decision points is unaffected by the flow. On the other hand, the number of decision points is affected by the number of barriers (particles) so that halving the size of the particles in a column while keeping the column volume the same would double the number of decision points. Concomitantly this would also halve the distance between decisions. This is an example of a random walk problem for which probability theory provides a relationship between σ_l, the standard deviation in the lengths L of the total walk, n, the number of steps (decisions) in the walk, and l, the length of each step, viz.,

$$\sigma = ln^{1/2} \tag{2.34}$$

Assuming that the length of each step can be approximately given as the diameter of the particle, it follows that the number of steps, n, is simply L/d_p. It follows that

$$\sigma = d_p(L/d_p)^{1/2} = (Ld_p)^{1/2} \tag{2.35}$$

Thus reducing the particle size of the packing will diminish the amount of band broadening, all other things being equal.

No real column has the perfection of packing implied by Figure 2.16, and superimposed on dispersion due to eddy diffusion is the usually more serious broadening caused by pathways of differing diameter, length, and therefore flow resistance. A packing of nonuniform particle size—normal for chromatography—and imperfections in the way the column is packed lead to pathways of essentially different bore diameter. When it is realized that the laws of fluid flow predict a fourth power dependence of volumetric flow on bore diameter it is not surprising that packing imperfections are a major contributor to band broadening in chromatography. The proper packing of columns is a major consideration in chromatography and unfortunately still something of an art. Snyder and Kirkland[9] give a useful treatment of some of the measures and recipes to increase the chances of success in this somewhat empirical area of chromatography. But choosing small particles of very narrow particle size distribution is clearly in the right direction.

It is appropriate to note that a step in a beneficial direction often incurs a penalty. In this case, a reduction in particle size leads to an increase in column pressure; the pressure drop is in fact inversely proportional to the square of the particle size. Choosing the appropriate physical parameters of a column therefore involves compromises between benefits and penalties.

Mass Transfer Resistance. While the results of the first two causes of band broadening are intuitively fairly obvious, mass transfer effects present more difficulty. It is not readily apparent how the rate of mass transfer relates to broadening or to the shape of the band. As in many problems it is often helpful to compare the real system to an ideal system. The ideal, hypothetical system chosen is one wherein equilibration is "infinitely" fast. As the mobile phase advances in the column there is a corresponding infinitely rapid shuttling of solute between the two phases, with a molecule of solute spending a short time $\delta\tau_m$ in the mobile phase and a correspondingly short time $\delta\tau_s$ in the stationary phase with the constraint that $\delta\tau_s/\delta\tau_m = k'$, satisfying the thermodynamic requirements. Thus as the solvent front advances even but a very small distance there have already been a great many transfers of a molecule between the phases. This will lead to a strong averaging effect so that all molecules will tend to elute with the same k-prime—a very sharp band in other words.

In real systems, however, owing to the sluggishness of transfer, molecules spend a finite time in each phase. This process can also be treated as a random walk process.[7] As the molecule makes its way through the column it is confronted frequently with the decision whether to be in the mobile or the stationary phase. The frequency of the choice will depend on the average time that the molecule spends in each phase, so that rapid exchange is consistent with more decisions, that is, more steps in the random walk process. Time spent in the stationary phase contributes nothing to solute advancement and therefore corresponds to a falling behind the center of the zone. How much the molecule falls behind depends on the time τ_s it spends in the stationary phase and the velocity of the mobile phase. Correspondingly if the molecule "chooses" the mobile phase it will get ahead of the zone center by a like amount. In fact the problem has a lot in common with the partitioning in the multistage contactor where now it is not length but time that is being subdivided into increments. It would not be surprising therefore if the mathematics yielded a similar result.

If the length of the column is denoted as L and the mobile phase velocity as U then it will take a time L/pU for the zone center to elute if p represents the fraction of time that the solute spends in the mobile phase. Recall also that $(1 - p)/p = k'$. During this time the molecule will spend a fraction $(1 - p)$ of its time in the stationary phase and if the average interval that it spends in the stationary phase is τ_s then the number of "desorptions" is simply

$$L(1 - p)/pU\tau_s \tag{2.36}$$

The number of steps made as it progresses through the column will be exactly twice this since each "desorption" has a corresponding number of sorptions. Thus the number of steps in the random walk process is

$$2L(1 - p)/pU\tau_s \tag{2.37}$$

What is the length, l, of the step? The length of the step is the distance the molecule falls back with respect to the band center during its relaxation time τ_s in the stationary phase. Since the band center is moving with velocity pU the step distance is therefore $pU\tau_s$.

If on the other hand the molecule chooses the mobile phase it will remain there for a time τ_m and the distance traveled will be $U\tau_m$. During this time the zone center travels a distance $pU\tau_m$ so the amount the molecule gets ahead of the zone center is $U\tau_m - pU\tau_m = U\tau_m(1 - p)$. At first glance it seems that the distance the molecule falls back while in the stationary phase, $pU\tau_s$, and the distances it gets ahead while in the mobile phase, $U\tau_m(1 - p)$, are different; an intuitive concern for symmetry, on the other hand, suggests that they should be the same. Recall, however, that $\tau_m/(\tau_m + \tau_s) = p$, from which it is easily shown that $\tau_m(1 - p) = p\tau_s$, which makes the two distances the same.

We now have the length of the step and the number of steps so, substituting in (2.34) gives

$$\sigma = [2p(1 - p)LU\tau_s]^{1/2} \tag{2.38}$$

This expression for the dispersion is consistent with the experimental observation that band spreading increases with flow rate and with any measure that tends to increase the relaxation time in the stationary phase such as slow diffusion throughout the particles of the packing. And the dependence on p (or k') indicates how band spreading is related to the retention of the solute.

The effects of longitudinal and eddy diffusion and mass transfer can be brought together in a single relationship since probability theory states that total variance in a system of several variances is given by a simple summation,

$$\sigma_{total}^2 = \sigma_L^2 + \sigma_E^2 + \sigma_M^2 \tag{2.39}$$

where the subscripts L, E, and M refer to longitudinal, eddy, and mass transfer, respectively; thus we have

$$\sigma_{total}^2 = (2DL/U) + Ld_p + 2p(1 - p)LU\tau_s \tag{2.40}$$

Now from (2.31) HETP $= \sigma^2/L$, so substituting in (2.40) gives

$$\text{HETP} = (2D/U) + d_p + 2p(1 - p)U\tau_s \tag{2.41}$$

This simply derived expression summarizes in an approximate way how band broadening depends on the various dynamic (kinetic) properties of a liquid chromatographic system. In particular, it expresses the dependence of column efficiency (HETP) on mobile phase velocity. Other expressions have been developed in a more empirical way from experimental observations of chromatographic efficiency. One common one is

$$\text{HETP} = B/U + AU^{0.33} + CU \tag{2.42}$$

where U is the mobile phase velocity and A, B, and C are constants.[9]

In this expression, as in (2.41), the terms on the right-hand side apply to the dispersions caused by, respectively, longitudinal, eddy, and mass transfer effects. Apart from the second term, where experimental data support a dependence of eddy diffusion on mobile phase velocity, the complementarity between the theoretically derived relationship and the one that is experimentally derived is evident.

To the practicing chromatographer such relationships have diagnostic value. By examining the effect of mobile phase velocity on the band broadening of an eluite, we can determine which factor, if any, is the dominant cause of band broadening for a particular eluite. In assessing the origin of band broadening, the chromatographer should be mindful that elements outside the chromatographic column also contribute to band broadening. Detectors and connecting tubing are the most likely sources.

2.6. RESOLUTION

The purpose of chromatography is to separate analytes into discrete bands. The quality of the separation in turn depends on the extent to which band centers move apart and on the degree of band dispersion. Resolution is the term most often used to describe the quality of a chromatographic separation. From measurements of peak elution volumes (times) and bandwidths one may calculate a value that expresses in a quantitative way the degree of resolution. An expression commonly used is

$$R = 2(V_2 - V_1)/(W_1 + W_2) \tag{2.43}$$

where R is resolution, V_1 and V_2 are the elution volumes of two adjacent eluites, and W_1 and W_2 are the peak widths at the base as determined by drawing tangents through the points of inflection of the peaks (Figure 2.17).

The resolution in a chromatographic separation is the result of two opposing effects: equilibrium or thermodynamic effects causing eluite bands to move apart, and kinetic effects that cause bands to overlap. Much of chromatographic art involves the optimization of these competing effects.

How should the chromatographer go about this optimizing process? Addressing the equilibrium aspects of mobile phase, stationary phase, and eluite is a good place to start. For example, one can often realize a significant improvement in overall resolution through a simple change in the composition of the mobile phase. As we will see later, this is especially true in ion chromatography.

The extent to which the chromatographer can intervene in the kinetic control of resolution depends largely on how much he is involved in the preparation of the chromatographic column. The manufacturer of columns will obviously be

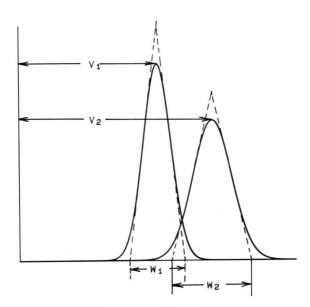

FIG. 2.17. Resolution.

concerned with such matters as the particle size of the stationary phase and how it is packed in the column, the permeability of the packing to likely eluites, and any other matters that impinge on the efficiency of the column. The analytical chemist who must usually accept what columns are available will have less latitude for intervention of this sort.

2.7. SUMMARY

Separation in chromatography is a result of the selective partitioning of solutes between a mobile and a stationary phase. The process is a dynamic one in that species are constantly transferring between phases as well as advancing through the chromatographic column under the influence of mobile phase flow. The thermodynamic aspects of the process may be conveniently approached by way of an equilibrium model where the relative affinities of species for the stationary phase are expressed in terms of a distribution coefficient K_D and the rate of movement of band centers is simply related to K_D. This preliminary step of considering the chromatographic process from a more static viewpoint is extremely helpful when complex equilibria between various forms of the eluite are involved. This is a common circumstance in ion chromatography, so future chapters will continue to develop an understanding of static equilibria as an effective means of understanding the more complex dynamic experiment.

The quality of chromatographic separation also depends on physical features of the system. Separation is enhanced by using stationary phases that present the shortest possible diffusional pathways to the eluites, have low resistance to mass transfer, are of reasonably narrow particle size distribution, and are uniformly packed in the column. As a rule, separation is improved by using lower mobile phase velocity.

REFERENCES

1. H. C. Subba Rao and M. M. David, Equilibrium in the System $Cu^{2+}-Na^+-Dowex$ 50, *AIChE J.* **3**, 187–191 (1957).
2. A. J. P. Martin and R. L. M. Synge, A New Form of Chromatogram Employing Two Liquid Phases. 1. A Theory of Chromatography. 2. Application to the Micro-determination of the Higher Monoamino Acids in Proteins, *Biochem. J.* **35**, 1358–1368 (1941).
3. H. Cramer, *The Elements of Probability Theory*. John Wiley and Sons, New York (1955).
4. J. J. van Deemter, F. J. Zuiderweg, and A. Klinkenberg, Longitudinal Diffusion and Resistance to Mass Transfer as Causes of Nonideality in Chromatography, *Chem. Eng. Sci.* **5**, 271–289 (1956).
5. H. Eyring and J. C. Giddings, A Molecular Dynamic Theory of Chromatography, *J. Phys. Chem.* **59**, 416–421 (1955).
6. J. C. Giddings, Stochastic Considerations of Chromatographical Dispersion, *J. Chem. Phys.* **26**, 169–173 (1957).

7. J. C. Giddings, Theory of Chromatography, in *Chromatography* (E. Heftmann, ed.), pp. 20–32, Reinhold Publishing, New York (1961).
8. J. Crank, *The Mathematics of Diffusion*. Clarendon Press, Oxford (1976).
9. L. R. Snyder and J. J. Kirkland, *Introduction to Modern Liquid Chromatography*, 2nd Ed. John Wiley and Sons, New York (1979).

Chapter 3

The Materials of Ion Chromatography

3.1. INTRODUCTION

Of all the elements in a chromatographic system, the stationary phase is the key. It is the stationary phase that determines what separation mechanism is operative, and that in turn dictates the choice and composition of the mobile phase and often what detection methods may be appropriately applied. To understand the function of this important element of the chromatographic process it is necessary to know something of its composition, and in this regard it is helpful to know how stationary phases are prepared. To these ends this chapter is devoted to the preparation, composition, and some of the basic properties of the materials that make up the stationary phase of ion chromatography.

3.2. ION EXCHANGERS

Ion exchangers are the most widely used stationary phase in IC. An ion exchanger comprises three important elements: an insoluble matrix, which may be organic or inorganic; fixed ionic sites, either attached to or an integral part of the matrix; and, associated with these fixed sites, an equivalent amount of ions of charge opposite to that of the fixed sites. The attached groups are often referred to as functional groups. The associated ions are called the counterions. They are mobile throughout the ion exchanger and most importantly have the ability to exchange with others of like charge when placed in contact with a solution containing such. It is this latter property that gives these materials their name.

As well as having this fundamental property, ion exchangers, if they are to be useful in IC, should also have the following properties:

1. The ability to exchange their ions rapidly.
2. Good chemical stability over a wide pH range.
3. Good mechanical strength and resistance to osmotic shock.
4. Resistance to deformation when packed in a column and subjected to the flow of the mobile phase.

A variety of materials have been used as matrices on which to anchor the ionic sites of an ion exchanger. In modern IC the two most widely used are silica and organic polymers based on styrene. The excellent chemical stability of the organic-based ion exchange resins gives them a distinct advantage over the pH-sensitive, silica-based materials in many applications.

3.2.1. Polymer Precursors of Ion Exchange Resins

The most common way of preparing an ion exchange resin is to first form a neutral, styrene-based polymer and then to chemically modify the polymer so as to introduce the ionic functional groups. Alternatively, monomers bearing the required functionality may be polymerized along with a suitable cross-linker to yield an ion exchanger, but materials prepared by this route are not in common use in IC.

There are two principal routes to the polystyrene precursor of most ion exchange resins. The first yields what are usually known as gel-type polymers, while the second yields macroporous materials. The difference in structure between the two types has important implications for ion exchange, particularly with regard to the rate of exchange of ions, especially when they are large.

Gel-type polymers are formed by copolymerizing styrene and divinyl benzene in the presence of a catalyst such as benzoyl peroxide. Divinyl benzene cross-links the polymer into a structure that will swell in solvents but will not dissolve in them. The degree of swelling of the polymer and of the ultimate ion exchange resin is usually controlled by the amount of DVB in the styrene–DVB monomer mixture, although in some cases subsequent reactions of the polymer can form additional cross-links. Cross-linking is indicated as a percentage, the mole percent of DVB in the monomer mixture.

Polymer particles are prepared in the form of small spherical beads by suspension polymerization techniques. The monomers with the dissolved catalyst are dispersed by agitation into small droplets in a water-based continuous phase and the mixture then heated to effect polymerization into solid particles. Suspending and dispersing agents added to the water help to control the size of the final beads and prevent their coalescence during polymerization. Beads as small as several micrometers in diameter may be prepared by suspension polymerization, but sizes much below this must usually be prepared by other means. Sub-micron-sized polymer spheres can be readily prepared by emulsion polymerization methods and, as we will see later, are an important precursor for IC resins.

Gel-type polymers and the ion exchange resins derived from them are often described as being microporous. The ion exchange resins will readily admit small ions and molecules but resist the intrusion of species of even a few hundred molecular weight. Diffusion through the particles is therefore improved by lowering the degree of cross-linking. Practical resins in IC are formed from polymers with anywhere from 1% to about 12% DVB. When the beads are fully functionalized, resins of low cross-linking are quite gelatinous and as a result are too deformable under flow to be broadly useful. So the limits of cross-linking on these type of resins are in the range of 4%–12%. The lower levels of cross-linking are used in those cases where the polymer is derivatized only on a thin outer shell. In the final resin it is only this shell that is swollen, the core remaining unswollen and rigid. Since the outer shell is only a small fraction of the total volume of the bead, a high degree of swelling in the shell can be tolerated.

Macroporous polymers are also formed by the suspension polymerization of styrene and divinyl benzene. In this case, however, another water-immiscible solvent is added that is miscible with the monomers but is a relatively poor solvent for the polymer. This added solvent is sometimes called the porogen since it is largely responsible for the macroporosity that is developed. As polymer forms, owing to its poor solubility in the porogen, it precipitates as a separate phase in the polymerizing droplet and the residual space is occupied by the porogen. The result is a bead with a high internal surface area permeated with rather large pores. A good way of picturing the final polymer bead is as a loose aggregate of very small sub-micron-sized particles with the interstitial volume between the particles constituting the porosity (Figure 3.1). Polymers of this type are prepared with large percentages of DVB so the resin material itself is quite rigid.

Macroporous polymers are widely used to prepare ion exchange resins for large-scale applications. In IC, however, they are mainly used in conjunction with ion interaction reagents (see Section 3.3 and Chapter 4) in techniques such as paired ion chromatography and mobile phase ion chromatography. For a more detailed background on the preparation, structure, and morphology of styrene-based polymers the reader is referred to a review by Guyot and Bartholin.[1]

3.2.2. Cation Exchange Resins

The most widely known and used cation exchange resin is formed by sulfonation of styrene-based polymers. Sulfonation, in the presence of a swelling agent, is a very efficient reaction in that essentially all of the aromatic rings in a low to moderately cross-linked polymer can be monosubstituted with $-SO_3^-$

A   B

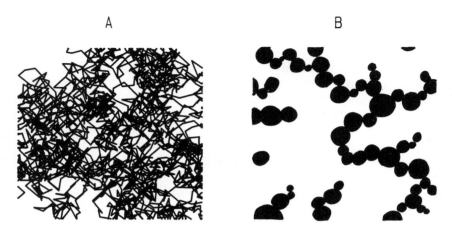

FIG. 3.1. Representation of a gel-type polymer (A) and a macroporous polymer (B). The gel consists of cross-linked and interlocking polymer chains that provide a network of very narrow pores that admit only small molecules or ions. The macroporous particle can be thought of as a loose aggregate of microparticles, 10–100 Å in diameter, for example. The interior of the macroporous particle is accessible by way of the relatively open interstitial space of the microparticle aggregate. Although the microparticles are highly cross-linked, the diffusional distance in these regions is small. Furthermore, in IC it is only the external surface of the microparticle that is used in most cases, so the overall rate of mass transfer is high.

groups. Resins that have a high degree of substitution with the requisite functionality are often termed ''conventional'' resins to distinguish them from the more specialized, low-capacity materials that are so widely used in IC. Capacity is the term used to define the concentration of charged sites and therefore exchangeable counterions in the ion exchanger. High-capacity cation exchangers have capacities of about 5 milliequivalents per dry gram. Their wet weight or wet volume capacity depends on the degree to which they are swollen by water. Table 3.1 shows how the swelling of sulfonated resins depends on the degree of cross-linking.

While high-capacity cation exchange resins may be used effectively as a stationary phase—amino acid separation was an early example of this—they do

TABLE 3.1. Weight Swelling of the Hydrogen Form of Sulfonated Polystyrene–DVB Resins[a]

Nominal percent of DVB	2	2.5	4.5	5	7.5	10	15	25
Weight swelling[b]	3.55	3.1	1.78	1.5	1.1	0.82	0.58	0.38

[a]Reference 2.
[b]Grams of water absorbed per gram of dry hydrogen form resin.

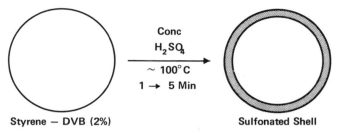

Useful Capacity About 0.02 meq/g

FIG. 3.2. Low-capacity cation exchange resins can be prepared by lightly sulfonating styrene–DVB copolymer particles.

have some drawbacks. They require rather concentrated eluents, which can complicate some detection methods (Chapter 7). Because of the low diffusivity of large or highly charged ions the resin particles must necessarily be made small in order to counteract this efficiency-lowering feature. This imposes a penalty of high pressure drop which will be aggravated if the cross-linking is low enough to allow deformation under flow. On the other hand, high-capacity resins do have the ability to handle large sample loads. They have two principal uses in modern IC, as the stationary phase in ion exclusion (Chapter 5) and in suppressor columns for anion analysis by suppressed conductometric detection (Chapter 7). This latter use is diminishing, however, as membrane-based suppressors take over.

Low-capacity cation exchange resins may be prepared by limiting the sulfonation of the polystyrene–DVB bead to a thin surface shell (Figure 3.2). This may be accomplished by eliminating the swelling agent and exposing the polymer, briefly, to hot concentrated sulfuric acid.[3] Sulfonation tends to progress into the bead with a relatively sharp boundary (see also Section 4.3.1), and the depth of substitution may be controlled by the time of contact, the temperature of the acid, and the cross-linking of the polymer. Stevens and Small[4] reported on the optimization of the surface sulfonation process. They found that polymers of about 2% DVB gave the best resins for cation separations. A useful range of capacities for IC is from about 0.01 to about 0.05 meq/g. Horvath and co-workers[5,6] have used the term "superficial" ion exchanger for materials prepared by modifying a thin surface layer of a substrate material.

3.2.3. Anion Exchange Resins

Ion exchange resins with attached quaternary ammonium groups are the principal anion exchangers of ion chromatography. Resins of high capacity are prepared by a sequence of reactions in which the styrene–DVB polymer is first chloromethylated and the pendant —CH$_2$Cl groups that are introduced are then quaternized with a tertiary amine R$_1$R$_2$R$_3$N. High-capacity anion exchangers have been used in IC principally in suppressor columns for cation analysis, but this application too is declining as membrane suppressors replace the older column methods.

Low-capacity anion exchangers may be prepared by surface quaternization of fully chloromethylated gel-type bead. The products, however, give poor performance and it is difficult to control their capacity.

Using macroporous polymers as starting materials, Barron and Fritz[7] were able to prepare low-capacity anion exchangers with useful separation properties. As a first step, they replaced the usual chloromethylating agent, chloromethyl methyl ether, with an aqueous system of paraformaldehyde and concentrated hydrochloric acid. This measure has two principal advantages over the conventional chloromethylation procedure. In the first place it is safer. Chloromethyl methyl ether (CMME) contains small amounts of bis-chloromethyl ether, which is a known human carcinogen.[8] Consequently procedures and laboratories that use CMME are now strictly regulated in how they handle this material.[9] The method of Barron and Fritz avoids direct use of CMME; however, they caution that *in situ* generation of bis-chloromethyl ether in the HCl–formaldehyde mixtures warrants proper precautions against contact with the vapors from the reaction. The second advantage of this procedure has to do with limiting the penetration of the chloromethylation into the substrate. Haloethers are excellent swelling agents for styrene–DVB polymers, so it is virtually impossible to limit or control their penetration. In the aqueous-based system the polymer remains unswollen and chloromethylation tends to be confined to a thin surface layer. In turn, quaternization is likewise confined to a thin surface layer.

Surface modification of this type inevitably leads to a zone of gradation from functionalized to nonfunctionalized polymer. This can give rise to sluggish diffusion of eluite counterions that enter this region which can lead to a loss in column efficiency in a chromatographic use. A method of preparing low-capacity exchangers that assures an abrupt transition from functionalized to nonfunctionalized polymer is therefore very desirable. The surface agglomeration approach to resin preparation is one such method.

3.2.4. Surface Agglomeration Method for Preparing Low-Capacity Ion Exchangers

Surface agglomeration is a term used to describe the attachment of colloidal ion exchange resin of one charge to a much larger substrate particle of the opposite charge.[10,11] In a typical procedure for making a low-capacity anion exchange resin, beads of surface sulfonated styrene–DVB are contacted with a

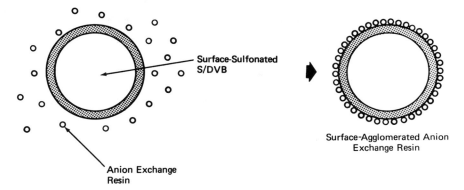

FIG. 3.3. Surface agglomeration method for preparing low-capacity anion exchange resins. Surface sulfonated styrene–DVB copolymer particles (5–50 μm diameter) are contacted with colloidal anion exchange particles (several hundred to several thousand angstroms in diameter).

suspension of colloidal anion exchange resin (Figure 3.3). The cationic functionality of the small particles is attracted to the anionic sulfonate groups on the larger bead and an extremely strong electrostatic attachment is established. Figure 3.4 is a schematic representation of this bonding; a more correct picture might be one that shows the polyelectrolyte chain end "fuzz" from one particle entangled and electrostatically attached at many points to its counterpart "fuzz" from the other particle.

In the early work on IC[10] the fine anion particles were produced by grinding larger particles of commercially available resins such as Dowex 1 and Amberlite IRA 400.* Subsequently Small and Solc[12] developed a more efficient and more controllable means of preparing the colloidal anion exchanger.

The first step involved the emulsion copolymerization of vinyl benzyl chloride (VBC) and DVB using a surfactant as emulsifier and a catalyst such as potassium persulfate (Figure 3.5). The size of the final polymer particles is controlled mainly by the amount of surfactant used. Large particles require a multistep procedure wherein small poly-VBC particles are used as "seed" on which to grow the larger particles.

After polymerization is complete, the poly-VBC latex emulsion is reacted with amine to form quaternary ammonium functionality on the latex particle. A slight stoichiometric excess of amine is sufficient in most instances to assure complete reaction of the chloromethyl groups in the latex particles.

*Amberlite is a trademark of the Rohm and Haas Company.

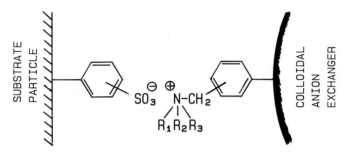

FIG. 3.4. In surface agglomerated anion exchange resins, colloidal anion exchange particles are held on the surface of the substrate beads by electrostatic interaction of sulfonate groups on this surface with the oppositely charged quaternary ammonium groups on the surface of the colloidal particle.

Surface agglomeration affords relatively simple control over several important properties of the final resin. The capacity may be controlled in essentially three ways: by the size of the substrate particles, by the size of the colloid, and by the extent of coverage of the substrate by the colloidal resin.

Early resins used substrates of about 20 μm in diameter and colloidal resin of 1000 Å or greater. Present-day resins employ substrates of less than 10 μm in diameter and colloidal resin of as low as a few hundred angstroms in diameter. Figure 3.6 is a scanning electron micrograph of a particle of surface agglomerated resin.

Efficiency is also affected by the size of both the colloid and the substrate. Stevens and Langhorst[13] found a twofold improvement in column efficiency over the early IC resins by employing microparticles of 200–1000 Å on a 15-μm substrate.

Work at Dionex Corporation has expanded the variety of resins that may be produced by this method and users of IC now have a wide range of columns of capacity, efficiency, and selectivity that may be matched to the application demands.

Slingsby and Pohl[14] have made an extensive study of how elution behavior of eluites is influenced by the nature of the amine used in the quaternization step. Barron and Fritz[15] have made similar studies on surface quaternized resins.

Several of the advantages of the surface agglomerization route to IC resins may be summarized as follows:

1. Column capacity can be controlled through choice of the size of the colloidal resin and of the substrate particles.

2. The degree of cross-linking of the colloidal resin can be controlled. This

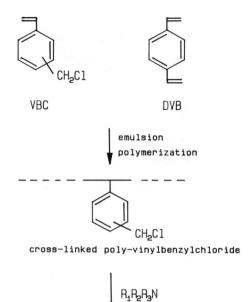

VBC   DVB

emulsion
polymerization

cross-linked poly-vinylbenzylchloride

$R_1R_2R_3N$

FIG. 3.5. Emulsion polymerization process for producing colloidal anion exchange resin particles.

has an important bearing on both resin selectivity and column efficiency. Resins based on surface modification of polymers, particularly macroporous polymers, are much more limited in this respect.

3. Resin properties have excellent reproducibility. One outstanding advantage of the emulsion process is the very large quantity of surface agglomerated resin that may be prepared from a modest amount of the colloidal resin. This is a boon to maintaining reproducibility in the resin manufacturing process.

4. There are no partially functionalized regions that can lead to sluggish mass transfer and poor efficiency. The layer of "active" resin has a uniform degree of solvation throughout its thickness, in contrast to a surface-modified polymer, which is likely to have poorly solvated regions. This offers such a clear advantage over materials prepared by surface modification procedures that Dionex workers have devised means of preparing low-capacity cation exchangers by the surface agglomeration technique.

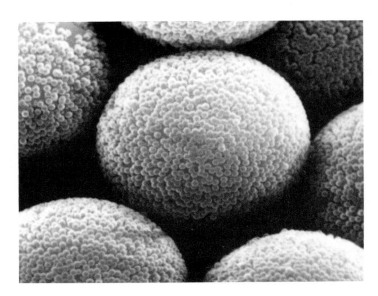

FIG. 3.6. Scanning electron micrograph of surface agglomerated anion exchange resin particles. The diameter of the substrate particles is about 20 μm and of the colloidal particles about 3000 Å. (Micrograph was prepared by E. Bradford of the Physical Research Laboratory of the Dow Chemical Company.)

5. Surface agglomeration allows increased safety. Emulsion polymerization of VBC yields the versatile chloromethylated polymer while avoiding the potential health hazards of chloromethylation reagents.

Horvath, who pioneered the introduction of low-capacity materials in liquid chromatography, described ion exchangers of the surface agglomerated type as "pellicular" resins; he used the term "superficial" ion exchange resins for materials formed by the surface modification of a preformed polymer. Among the advantages Horvath cites for pellicular resins are the following:

1. Pellicular resins have the ability to withstand pressure gradients without changes in packing structure. Horvath was primarily referring to exchangers wherein silica was used as substrate for the pellicle of active material, but the same holds true for resins that use polystyrene–DVB substrates. Because only a thin surface layer is reacted to provide an anchor for the colloidal resin, the core remains unswollen and the composite resin is therefore relatively rigid and will not deform significantly under the usual chromatographic flow rates.

2. Pellicular resins have stable volume when exposed to extremes of eluent composition. It is assumed that for this to be true we are dealing with eluent

components to which the substrate particles are inert. For polystyrene substrates this would not necessarily apply when organic solvents are part of the eluent, particularly if they are good swelling agents for these polymers. A great many IC methods do not require the use of solvents other than water, and in this environment surface agglomerated resins and columns prepared from them are stable.

3. Pellicular structures such as surface agglomerated resins permit the use of slightly cross-linked resins in the pellicle layer. This extends the applicability of ion exchange resins to higher molecular weight ions. Conventional resins of high capacity and low cross-linking are simply too deformable to be of practical use at reasonable flow rates.

3.2.5. Silica- and Methacrylate-Based Ion Exchangers

Low-capacity ion exchangers built on a silica base have been known for some time.[16] They have been prepared by bonding ionically modified polymers to a totally porous silica microparticle or to a superficially porous particle. Both sulfonate and quaternary ammonium type exchangers are available. Exchangers of this type have good mass transfer characteristics since their ion exchange sites are accessible via relatively large pores. Not surprisingly, therefore, silica-based ion exchangers have provided excellent separations of large organic ions, especially those of biological origin such as polypeptides, vitamins and nucleotides. The book by Done, Knox and Loheac[17] provides a good many examples of the chromatographic applications of these earlier silica-based exchangers.

For the chromatography of small ions, organic as well as inorganic, exchangers based on silica have a number of disadvantages relative to the totally organic ion exchange resins. Their main drawback is their notorious instability in some of the aqueous eluent systems that are so typical of much of modern IC. This is particularly so in high pH eluents. Additionally, the chemistry of bonding to silica does not afford the flexibility in such variables as capacity, cross-linking, and selectivity that is the case in polymer systems. For these reasons, silica-based ion exchangers have a limited use in the IC of small organic and inorganic ions.

Acrylate-based ion exchangers manufactured by the Toya Soda company are being increasingly used in IC, although details on their preparation and structure are sparse. In the basic environments common to many IC systems these resins are superior to silica-based materials but are less stable than resins based on polystyrene. Some anion exchangers manufactured by Dionex use an acrylate latex to form the pellicular layer. These materials have an advantage over polystyrene-based exchangers in that peak shape and retention time of solutes are much less susceptible to change as the amount of solute is increased.[18] They are, in other words, less prone to overload.

3.2.6. Exchangers with Weakly Functional Groups

Most of the ion exchangers used in modern IC contain strongly acidic or basic groups. Such groups by definition remain highly ionized over a broad range of pH. Some exchangers, on the other hand, contain weakly functional groups whose ion exchange activity is dependent on the pH of the environment. Materials containing the carboxylic acid functional group exemplify weak cation exchangers, while primary, secondary, and tertiary amino substituents are the principal functional groups of weak base resins.

For weak acid resins, their capacity increases with increasing pH, while the reverse is the case for weak base resins. The ability to control ion exchange capacity by varying the pH of the environment can be a very effective means of modulating the overall affinity of the ion exchanger for the eluite ions. However, compared to the use of strongly functional materials, there has been limited application of resins with weak functionality.[19,20]

3.2.7. Chelating Resins

Since the introduction of organic ion exchange resins there has been a steady interest in preparing polymers containing groups with the ability to complex or chelate metal ions.[21-23] Ion exchange resins with chelating groups have been examined from time to time as stationary phases in the separation of metal ions. While these resins have a greater selectivity than conventional exchangers, they do not enjoy widespread use as a chromatographic medium. There are two probable reasons for this. Chelating resins as a rule tend to be kinetically inferior to conventional resins, and the resulting inefficiency somewhat cancels their superior selectivity. But perhaps a more compelling reason for their lack of popularity is that conventional ion exchangers used in conjunction with chelating or complexing agents in the mobile phase is a very powerful means of metal ion separation (Section 4.2.7). The variety, availability, and relatively low cost of chelating and complexing agents, and the flexibility in the choice of mobile phase composition, makes for an extremely versatile separation method. To this add the fact that use of a chelating resin does not afford any significant advantages with regard to detection and it is apparent why methods using conventional resins are preferred.

Chelating resins do, however, have an important place in the total ion analysis scheme. Their great selectivity range provides for a very effective means of concentrating certain metal ions from high background levels of competing ions, a task for which the conventional resins are ill suited. There are a number of commercially available resins that are variants on the first commercial chelating resin—Dowex A-1. They contain the iminodiacetate functional group and display a high affinity for many important metal ions relative to such common cations as the alkali metal and alkaline earth ions (Table 3.2). They have excel-

TABLE 3.2. Selectivity of a Commercial Chelating Resin, Dowex A-1, for Metal Ions Relative to Zn^{2+} [a]

M^{2+}	K_{Zn}^{M} [b]
Hg	1060
Cu	126
Ni	4.4
Zn	1
Co	0.61
Mn	0.024
Mg	0.009

[a] Reference 24.
[b] The selectivity coefficient, K_{Zn}^{M}, was determined from equilibrium measurements on the distribution of radio-labeled zinc (Zn^{65}) between Dowex A-1 and a 0.1 M solution of $M(NO_3)_2$. K_{Zn}^{M} is the equilibrium coefficient for the reaction

$$Zn_R + M_S \rightleftharpoons M_R + Zn_S$$

where the subscripts R and S refer to the resin and solution phases, respectively.

lent ability to concentrate metal ions out of high-salt environments such as brines and seawater.

3.3. ION INTERACTION REAGENTS

An important class of IC methods uses a neutral nonpolar stationary phase in conjunction with mobile phases containing a lipophilic electrolyte—the ion interaction reagent (IIR). Many different names have been applied to this technique: soap form,[25] ion-pair,[17,26] dynamic ion exchange,[27–29] and ion interaction chromatography[30] are some of them.

Typical of the stationary phases used are porous silica whose surface has been modified by attaching long alkyl groups (C_8 or C_{18}, for example), and macroporous styrene–DVB polymers. The silica-based materials are also extensively used in reverse phase liquid chromatography.

For cation analysis on polymeric stationary phases, octane sulfonic acid is a typical IIR, while tetrabutyl ammonium hydroxide is commonly used for anion separations. An exact mechanism for this type of system is probably some way off, but one popular model builds on the analogy between the covalently bound functional group of a conventional ion exchanger and the sorptively attached lipophilic reagent.

Ion interaction chromatography has several advantages over ion exchange chromatography in some applications. A number derive from the ability to modulate the "capacity" of the pseudoexchanger simply by altering one or more of the following:

1. The concentration of the IIR in the mobile phase.
2. The lipophilicity of the IIR—in a homologous series of IIRs the longer the nonpolar chain the more the IIR is adsorbed by the substrate.
3. The solvent polarity of the mobile phase—the addition of water-miscible solvents such as alcohols or acetonitrile will diminish the interaction between the lipophile and the nonpolar stationary phase.

This ability to raise and lower capacity can be a boon in dealing with samples that contain ions with a wide range of affinities. Furthermore, since large pore substrates are used and the functionality of the pseudoexchanger is readily accessible on the pore walls, IIR methods are especially suitable for the chromatography of large ions. A fuller account of the mechanism of operation and applications of ion interaction reagents is given in Chapter 4.

REFERENCES

1. A. Guyot and M. Bartholin, Design and Properties of Polymers as Materials for Fine Chemistry, *Prog. Polym. Sci.* **8**, 277–332 (1982).
2. K. W. Pepper, D. Reichenberg, and D. K. Hale, Properties of Ion-Exchange Resins in Relation to Their Structure. Pt. IV. Swelling and Shrinkage of Sulphonated Polystyrenes of Different Cross-Linking, *J. Chem. Soc.* 3129–3136 (1952).
3. H. Small, Gel Liquid Extraction. The Extraction and Separation of Some Metal Salts Using Tri-*n*-Butyl Phosphate Gels, *J. Inorg. Nucl. Chem.* **18**, 232–244 (1961).
4. T. S. Stevens and H. Small, Surface Sulfonated Styrene Divinyl Benzene. Optimization of Performance in Ion Chromatography, *J. Liq. Chromatogr.* **1**(2), 123–132 (1978).
5. C. G. Horvath, B. A. Press, and S. R. Lipsky, Fast Liquid Chromatography. An Investigation of Operating Parameters and the Separation of Nucleotides on Pellicular Ion Exchangers, *Anal. Chem.* **39**, 1422–1428 (1967).
6. C. G. Horvath, Pellicular Ion Exchange Resins in Chromatography, in *Ion Exchange and Solvent Extraction*, Vol. 5 (J. A. Marinsky and Y. Marcus, eds.), Marcel Dekker, New York (1973).
7. R. E. Barron and J. S. Fritz, Reproducible Preparation of Low-Capacity Anion-Exchange Resins, *React. Polym.* **1**(3), 215–226 (1983).
8. W. G. Pequeroa, R. Raszkowski, and W. Weiss, Lung Cancer in CMME Workers, *New Engl. J. Med.* 1096–1097 (1973).
9. OSHA Regulation Part 29CFR 1910.1200 and OSHA Regulation Part 29 CFR 1910.1008.
10. H. Small, T. S. Stevens, and W. C. Bauman, Novel Ion Exchange Chromatographic Method Using Conductimetric Detection, *Anal. Chem.* **47**, 1801–1809 (1975).
11. H. Small and T. S. Stevens, High Performance Ion Exchange Composition. U.S. Patent No. 4,101,460 (1978).

12. H. Small and J. Solc, Ion Chromatography—Principles and Applications, in *The Theory and Practice of Ion Exchange* (M. Streat, ed.), The Society of Chemical Industry, London (1976).

13. T. S. Stevens and M. Langhorst, Agglomerated Pellicular Anion-Exchange Columns for Ion Chromatography, *Anal. Chem.* **54,** 950–953 (1982).

14. R. W. Slingsby and C. A. Pohl, Anion-Exchange Selectivity in Latex-Based Columns for Ion Chromatography, *J. Chrom.* **458,** 241–253 (1988).

15. R. E. Barron and J. S. Fritz, Effect of Functional Group Structure and Exchange Capacity on the Selectivity of Anion Exchangers for Divalent Anions, *J. Chromatogr.* **316,** 201–210 (1984).

16. L. R. Snyder and J. J. Kirkland, *Introduction to Modern Liquid Chromatography,* 2nd Ed. John Wiley and Sons, New York (1979).

17. J. N. Done, J. H. Knox, and J. Loheac, *Applications of High-Speed Liquid Chromatography.* John Wiley and Sons, New York (1974).

18. C. A. Pohl, private communication.

19. H. J. Cortes and T. S. Stevens, High-performance Liquid Chromatography of Non-UV Absorbing Anions Using Indirect Photometric Chromatography and an Amino Column, *J. Chromatogr.* **295,** 269–275 (1984).

20. G. Domazetis, Determination of Anions by Nonsuppressed Ion Chromatography Using an Amine Column, *Chromatographia* **18**(7), 383–386 (1984).

21. G. Nickless and G. R. Marshall, Polymeric Coordination Compounds. The Synthesis and Applications of Selective Ion-Exchangers and Polymeric Chelate Compounds, *Chromatogr. Rev.* **6,** 154–190 (1964).

22. R. M. Wheaton and M. J. Hatch, Synthesis of Ion-Exchange Resins, in *Ion Exchange and Solvent Extraction*, Vol. 2 (J. A. Marinsky and Y. Marcus, eds.), Marcel Dekker, New York (1969).

23. H. Small, Chemical Modification of Crosslinked Polymers, *Ind. Eng. Chem. Product Res. Dev.* **6**(3), 147–150 (1967).

24. H. Small, unpublished results.

25. J. H. Knox and G. R. Laird, Soap Chromatography—A New High-Performance Liquid Chromatographic Technique for Separation of Ionizable Materials, *J. Chromatogr.* **172,** 17–34 (1976).

26. J. H. Knox and J. Jurand, Separation of Catecholamines and their Metabolites by Adsorption, Ion-Pair and Soap Chromatography, *J. Chromatogr.* **125,** 89–101 (1976).

27. B. Fransson, K. G. Wahlund, I. M. Johansson, and G. Schill, Ion-Pair Chromatography of Acidic Drug Metabolites and Endogenic Compounds, *J. Chromatogr.* **125,** 327–344 (1976).

28. J. C. Kraak, K. M. Jonker, and J. F. K. Huber, Solvent-Generated Ion-Exchange Systems with Anionic Surfactants for Rapid Separations of Amino Acids, *J. Chromatogr.* **142,** 671–688 (1976).

29. R. M. Cassidy and S. Elchuk, Dynamically Coated Columns for the Separation of Metal Ions and Anions by Ion Chromatography, *Anal. Chem.* **54,** 1558–1563 (1982).

30. B. A. Bidlingmeyer, S. N. Deming, W. P. Price, B. Sachok, and M. Petrosek, Retention Mechanism for Reversed-Phase Ion-Pair Liquid Chromatography, *J. Chromatogr.* **186,** 419–434 (1979).

Chapter 4

Ion Exchange in Ion Chromatography

4.1. INTRODUCTION

Of the many methods that have been applied to the chromatographic separation of ions, most employ ion exchangers. The purpose of this chapter is to describe some of the more important phenomena associated with the ion exchange process, to explain their physical causes, and to demonstrate how principles and theories of ion exchange relate to the chromatographic separation of ions.

Since chromatography is as much a matter of detection of eluted species as it is of their separation, the problems of detection will be broached from time to time in order to establish and reinforce the linkage between these two important areas.

While modern IC uses mainly low-capacity materials and their chromatographic characteristics are widely reported, their design and the understanding of their chromatographic behavior is based in large part on experience with the older, high-capacity materials. Therefore it will be the usual practice in this chapter to discuss the principles and theories developed for high-capacity exchangers and on that basis derive adaptations suitable to the lower-capacity materials.

4.2. ION EXCHANGE SELECTIVITY AND EQUILIBRIA

4.2.1. Selectivity and the Selectivity Coefficient

In a typical IC experiment, an ion exchange resin is loaded in a column and converted to a counterion "form" that will be denoted as E. A sample containing several eluite species (denoted S_1, S_2, etc.) of the same sign of charge as E is then loaded to the resin and an exchange takes place that deposits the sample species on the resin and liberates an equivalent amount of E. The column is then washed with a solution containing E, the eluent ion, and the sample ions are displaced downwards in the column as they undergo further exchanges between the solution and resin phases. The displacement is a selective process, with species moving at different rates through the column, and it is this selectivity that is a key factor in the success of ion exchange as an ion separation method. We will now examine selectivity as it is manifest in the behavior of typical cation and anion exchange resins.

The cation exchanger most commonly used in IC is a strong acid resin derived from the sulfonation of styrene divinyl benzene copolymers. For the time being we will consider the ion exchange behavior of resins with a high degree of sulfonation and restrict the discussion of exchange reactions to those involving ions of equal valency.

When, for example, the water-swollen hydrogen form of such a resin is placed in contact with an aqueous solution containing sodium ions, an exchange will take place so that sodium ions enter the resin phase and an equivalent number of hydronium ions replace them in the aqueous phase. If the solution phase should contain a species that reacts strongly with hydronium ions, hydroxide ions, for example, and in an amount sufficient to react with all of the hydronium ions, then the resin will be converted completely to the sodium form and the reaction can be represented simply as

$$R-SO_3^- H^+ + NaOH \rightarrow R-SO_3^- Na^+ + H_2O \qquad (4.1)$$

Should the sodium be accompanied with a co-ion such as chloride that is not reactive with the hydronium ion, then the ion exchange will, after sufficient time, reach a position of equilibrium where some sodium and hydronium ions remain in the solution and resin phases, respectively. This reaction may be represented as follows:

$$R-SO_3^- H^+ + NaCl \rightleftharpoons R-SO_3^- Na^+ + HCl \qquad (4.2)$$

The reaction is reversible in that this position of equilibrium can be approached from either direction.

For reasons that will be explained in detail later (Chapter 5, Section 5.2) there is very little invasion of the resin phase by the anion in the solution phase, especially when the external phase is dilute (less than $0.1\ M$). So with the exception of the case where the anions react or complex with one or more of the cations in the system they may be considered as of somewhat incidental importance, their major role being that of maintaining electroneutrality in the external aqueous phase. This "noninvolvement" of the co-ions is implied in the commonly used alternative to expression (4.2) that omits anions:

$$R-SO_3^- H^+ + Na_s^+ \rightleftharpoons R-SO_3^- Na^+ + H_s^+ \qquad (4.3)$$

Similar expressions are common in describing anion exchange equilibria where in this case the cationic co-ions are frequently omitted.

The concentrations of the various ionic species in the two phases may be measured and the equilibrium expressed in the form of a mass action relationship as follows:

$$K_H^{Na} = [Na^+]_R[H^+]_S/[H^+]_R[Na^+]_S \qquad (4.4)$$

where K_H^{Na} is an **equilibrium coefficient** and the subscripts R and S refer to the resin and solution phases. Concentrations in the two phases may be expressed in any of several ways: molarity, molality, equivalent fraction, or mole fraction.

If ion exchange systems were totally ideal, ions would behave as equivalent point charges acting independently of each other, the position of equilibrium would be determined solely by the relative amounts of sodium and hydrogen, and K_H^{Na} would be unity. Not surprisingly, ion exchange systems are nonideal and selectivity coefficients differ from unity, often quite markedly. Ion exchangers, in other words, show a preference for one ion over the other and K_H^{Na} is often referred to as the **selectivity coefficient.**

This preference is evident from Table 4.1, which summarizes the selectivity coefficients for the exchange of several cations on sulfonated polystyrenes (Dowex 50) with respect to hydronium ion as reference. It is evident that for the alkali metal ions a resin of this type displays a preference in the order

$$Cs^+ > Rb^+ > K^+ > Na^+ > Li^+$$

The order for the alkaline earth ions is

$$Ba^{2+} > Sr^{2+} > Ca^{2+} > Mg^{2+}$$

Figure 4.1 illustrates an early IC separation of the alkali metal ions and it is evident that the order of elution reflects the order of selectivities.

Selectivity differences are usually larger the higher the cross-linkage of the

TABLE 4.1. Equilibrium Constants for the Exchange of Various Cations for H^+ on Sulfonated Polystyrene (Dowex 50) of 4% Cross-Linking[a]

Counterion (M)	K_H^M	Counterion (M)	K_H^M
H^+	1	Mg^{2+}	2.23
Li^+	0.76	Ca^{2+}	3.14
Na^+	1.2	Sr^{2+}	3.56
NH_4^+	1.44	Ba^{2+}	5.66
K^+	1.72	Co^{2+}	2.45
Rb^+	1.86	Ni^{2+}	2.61
Cs^+	2.02	Cu^{2+}	2.49
Ag^+	3.58	Zn^{2+}	2.37
Tl^+	5.08	Pb^{2+}	4.97
		UO_2^{2+}	1.79

[a]Reference 1. Copyright 1957 American Chemical Society.

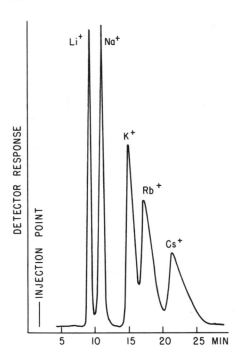

FIG. 4.1. Separation of alkali metals on a surface sulfonated styrene–DVB resin. (From Ref. 2 with permission.)

resin.[1] Furthermore, the selectivity coefficient reported is often an average value, for it depends not only on the ions involved and the cross-linkage of the resin but also on the composition of the resin phase at which the equilibrium is measured (Figure 4.2).

The principal anion exchanging resins of IC contain quaternary ammonium functionality. Selectivity data for two high-capacity versions of such resins are summarized in Table 4.2, and it is apparent that anion exchangers also display wide differences in their preference for their counterions. Like cation exchangers the equilibrium is affected by the cross-linkage of the resins and by the composition of the phases at equilibrium.

The magnitude of the equilibrium coefficient and its variations with resin composition reflect the nonideality of the system. A true thermodynamic equilibrium *constant* could be obtained if activity coefficients were available for the ionic species in both phases. Such data are available for the solution phase but not for the resin phase. However, in a typical ion chromatographic experiment where the eluent ion is usually in great excess over the eluite ions, a constant limiting value for the selectivity coefficient may be assumed.

Clearly ion exchange equilibria are complex and there has been much effort to account for the various effects. There are essentially two approaches to the

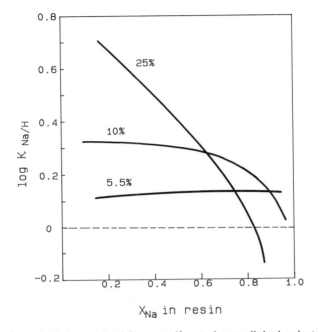

FIG. 4.2. Sodium–hydrogen selectivity on sulfonated cross-linked polystyrene resins showing the effect of resin composition and the degree of cross-linking. The numbers on the curves refer to the mole percentage of divinylbenzene in the copolymer from which the ion exchange resin was prepared. X_{Na} is the equivalent fraction of sodium in the resin at equilibrium. (From Ref. 3 with permission of Marcel Dekker, Inc.)

problem. One, the thermodynamic approach, attempts to describe the nonideality from independent measurements of properties of the homoionic resins, coupled with activity coefficient relationships for the two phases. The other approach attempts to relate ion selectivities and sequences to physical causes: electrostatic effects, solvation forces, hydrophobic interactions, and the like. This approach, though more qualitative and somewhat intuitive, is of more direct relevance to the practical side of ion chromatography. Models of this type when they have validity can be helpful in designing new ion exchange materials to solve specific separation problems. A later section (Section 4.2.8) is devoted to a discussion of some of these models.

4.2.2. The Distribution Coefficient K_D

While equations such as (4.4) are a common way of describing ion exchange equilibria, they are not directly applicable to chromatographic behavior. Chapter 2 showed how the chromatographic retention of a species may be con-

TABLE 4.2. Anion Exchange Equilibrium
Coefficients $(K)^a$

Anion	K (Dowex-1)	K (Dowex-2)
β-Naphthalene sulfonate		67
Dichlorophenate		53
Salicylate	32.2	28
Perchlorate		32
Thiocyanate		18.5
Trichloroacetate		18.2
p-Toluenesulfonate		13.7
Iodide	8.7	13.2
Phenate	5.2	8.7
Bisulfate	4.1	6.1
Benzene sulfonate		4.0
Nitrate	3.8	3.3
Bromide	2.8	2.3
Trifluoroacetate		3.1
Dichloroacetate		2.3
Nitrite	1.2	1.3
Bisulfite	1.3	1.3
Cyanide	1.6	1.3
Chloride	1.0	1.0
Bisilicate		1.13
Bromate		1.01
Hydroxide	0.09	0.65
Bicarbonate	0.32	0.53
Dihydrogen phosphate	0.25	0.34
Monochloroacetate		0.21
Iodate		0.21
Formate	0.22	0.22
Acetate	0.17	0.18
Fluoride	0.09	0.13

^{a}Reference 4.

veniently and effectively described by the distribution coefficient, K_D, or k-prime (k') and that differences in K_D or k' are the source of separation in chromatography. So in this chapter we will develop expressions for the K_D of ionic species that relate it to such important ion exchange properties and chromatographic variables as the following:

1. The selectivity coefficients of the various ion exchange reactions.
2. The capacity of the ion exchanger.
3. The concentration of the electrolyte in the mobile phase.
4. The charge of the eluent and eluite ions.

5. The pH of the mobile phase.
6. Complexation in the mobile phase.

Returning to the example of sodium and hydrogen exchange, an expression for the distribution coefficient of the sodium ion may be derived by a simple rearrangement of equation (4.4):

$$[Na^+]_R/[Na^+]_S = K_H^{Na}[H^+]_R/[H^+]_S \tag{4.5}$$

The term on the left-hand side is clearly the distribution coefficient, K_D, of the sodium ion between the resin and solution phases. It is usual in an IC experiment for one ion, the displacing or eluent ion, to be in much greater abundance than the eluite species. In cation IC the hydronium ion is typically the displacing species, so that the term $[H^+]_R$ approaches very closely the capacity of the resin, C_R, and (4.5) can be expressed as

$$[Na^+]_R/[Na^+]_S = K_D = K_H^{Na}C_R/[H^+]_S \tag{4.6}$$

The equilibrium coefficient, though it may depend strongly on the composition of the ions in the resin, tends to a constant limiting value as the fraction of sodium ion in the system approaches zero. So K_D tends to be independent of resin composition at low loadings of sodium.

Equation (4.6) is a particular form of a more general expression

$$[S]_R/[S]_S = K_D = K_E^S C_R/[E]_S \tag{4.7}$$

that describes the distribution of a monovalent counter ion, S, between an ion exchange resin and a solution in which the only competing monovalent counter ion is E, and E is greatly in excess of S. The symbols S and E refer to the sample ions and eluent ion of a typical chromatographic experiment. The omission of charge from the symbols implies that the expression is equally applicable to anion or cation exchange. Simple expressions of this sort have great utility in IC since they relate K_D directly to important parameters of the chromatographic system.

Most importantly the distribution coefficient is related to the preference the ion exchanger displays for E versus S as reflected in the selectivity coefficient. There are as yet no a priori means for calculating selectivity coefficients; practitioners of IC must either determine them or rely on the data of others. The same can be said for the sequences of ion affinities, although here qualitative arguments have been developed that in many instances appear to have validity. Some of this work, which has important predictive value for the design of ion exchange media, is discussed in Section 4.2.8.

The theoretical treatment so far has been confined to the exchange of monovalent ions. The exchange of ions of different valency may be illustrated by the exchange of hydrogen for calcium on a strong acid resin

$$2RSO_3^-H^+ + Ca_S^{2+} \rightleftharpoons (R-SO_3^-)_2\,Ca^{2+} + 2H_S^+ \qquad (4.8)$$

A mass action expression describing the equilibrium may be written

$$K_H^{Ca} = [Ca^{2+}]_R[H^+]_S^2/[H^+]_R^2[Ca^{2+}]_S \qquad (4.9)$$

which, through rearrangement and applying the assumption that calcium is in trace amount, yields the expression for the distribution coefficient for calcium:

$$[Ca^{2+}]_R/[Ca^{2+}]_S = K_D = K_H^{Ca}C_R^2/[H^+]_S^2 \qquad (4.10)$$

Since K_H^{Ca} and C_R are constants, this implies that the distribution coefficient for calcium is inversely proportional to the *square* of the hydronium ion concentration. (More correctly, it is inversely proportional to the square of the hydronium ion *activity,* but low concentrations will be assumed, where activity coefficients are approximately unity.)

This power dependence of K_D on solution concentration is more obviously conveyed in the plot of Figure 4.3, which shows K_D as a function of hydronium concentration for a typical monovalent and divalent ion. As the solution concentration is decreased the resin phase increasingly prefers the ion of higher

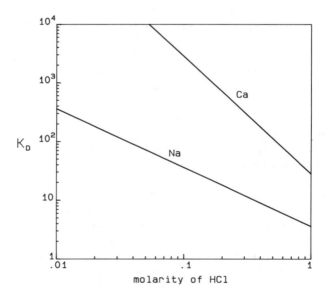

FIG. 4.3. Calculated distribution coefficients (K_D) as a function of acid concentration for the distribution of Na$^+$ and Ca^{2+} between hydrochloric acid and a fully sulfonated styrene–DVB resin. The capacity of the resin was assumed to be 3 meq/ml and selectivity coefficients were taken from Table 4.1. Concentrations were used in the calculation in place of activities. Note the very high values of the distribution coefficients for the subject ions from the most dilute solutions for this high-capacity resin.

charge. This is a well-known rule of ion exchange reactions that has important implications for ion chromatography.

Although the power dependence of K_D on eluent concentration (activity) is thermodynamically correct it is nevertheless not obvious why it should take that form. We will now derive the relationship in a somewhat different way that provides an intuitively more satisfying explanation and at the same time introduces an important concept in ion exchange—the Donnan potential.

4.2.3. The Donnan Potential and Ion Exchange Equilibrium

When a water-swollen ion exchange resin is placed in contact with a solution phase there is a natural tendency for all *mobile* species in the system to reach the same level in both phases. For example, when a hydrogen form resin is placed in a dilute solution of hydrochloric acid, diffusional forces try to level the large difference in hydronium ion concentration between the resin phase (typically about 3 molar) and the dilute solution phase. Significant loss of hydronium ions from the resin phase is prevented, however, by the immobility of the anionic sulfonate groups that are attached to the resin matrix. Electroneutrality must be preserved, or, expressed in another way, when hydronium ions leave the resin phase to disturb electroneutrality even minutely, any further loss of hydronium ions is opposed by the electrostatic potential that develops between the resin and solution. In the case of a cation exchanger the solution phase becomes positive with respect to the resin while the reverse is true for anion exchangers. Electrostatic considerations also show that equilibrium will be reached when an extremely small amount of mobile counterions leave the resin phase—in fact, an amount too small to measure. Thus the concentrations of hydronium ions in the resin and solution phases are almost exactly what they were before the phases were brought in contact, but a large potential difference now exists between the two phases. This is commonly called the Donnan potential. (The name derives from F. G. Donnan, who, at the beginning of this century, described the behavior of electrolyte solutions separated by membranes impermeable to one of the ions in the system[5]; modern workers recognized the parallels between Donnan systems and ion exchangers and applied the theory and adopted the name.)

We will now consider how this potential influences the distribution of various ions in the system. Clearly, while it favors the retention of counterions it also opposes the invasion of the resin by the co-ions in the system—chloride ions in this case—and equilibrium is reached when but a small amount of chloride enters the resin phase. This behavior is the basis of "ion exclusion" and is discussed more fully in Chapter 5.

How does the Donnan potential affect the distribution of cations, especially ions that have a different charge from the majority ion? The influence of electrostatic forces as well as mass action effects is conveyed in a more quantitative way in the following thermodynamic derivation.

At equilibrium, the chemical potential of an ionic species is the same in both phases. Thus for hydronium ions we have

$$\mu_0 + RT \ln\{H^+\}_R = \mu_0 + RT \ln\{H^+\}_S + Z_H F\psi \qquad (4.11)$$

where $\{\ \}$ denotes activities, Z_H is the charge on the hydronium ion, F is the Faraday constant, ψ is the Donnan potential, and μ_0 is the chemical potential of the hydronium ion in some standard reference state. The third term on the right-hand side represents the electrostatic contribution to the total free energy of the hydrogen ion in the solution phase. Rearrangement of (4.11) gives

$$\psi = (RT/Z_H F)\ln(\{H^+\}_R/\{H^+\}_S) \qquad (4.12)$$

If another ion is present in the system besides hydronium, sodium for example, the same argument yields

$$\psi = (RT/Z_{Na}F)\ln(\{Na^+\}_R/\{Na^+\}_S) \qquad (4.13)$$

If the sodium ion is present in trace amount it can be assumed that the Donnan potential has approximately the same value as in a pure hydronium system, so combining (4.12) and (4.13) and eliminating ψ gives

$$\ln(\{Na^+\}_R/\{Na^+\}_S) = (Z_{Na}/Z_H)\ln(\{H^+\}_R/\{H^+\}_S) \qquad (4.14)$$

Converting from logarithmic to an exponential expression gives

$$\{Na^+\}_R/\{Na^+\}_S = (\{H^+\}_R/\{H^+\}_S)^{Z_{Na}/Z_H} \qquad (4.15)$$

Since the activity of a species may be expressed as the product of its concentration and an activity coefficient, (4.15) may be expressed as

$$[Na^+]_R/[Na^+]_S = (K[H^+]_R/[H^+]_S)^{Z_{Na}/Z_H} \qquad (4.16)$$

where K represents the "lumping together" of activity coefficients. For dilute solutions and where one of the ions is at trace level, K can be assumed to be constant. When sodium is in trace amount, this leads to the further simplification

$$[Na^+]_R/[Na^+]_S = (KC_R/[H^+]_S)^{Z_{Na}/Z_H} \qquad (4.17)$$

Since sodium and hydronium ions have the same charge ($Z_{Na} = Z_H$), it is evident that (4.17) has exactly the same form as (4.6) where the constant K of (4.17) is the selectivity coefficient of (4.6) and the distribution coefficient has the same inverse first power dependence on solution concentration.

Pursuing a similar argument for trace calcium in a bulk of hydronium yields

$$[Ca^{2+}]_R/[Ca^{2+}]_S = (K'C_R/[H^+]_S)^{Z_{Ca}/Z_H} \qquad (4.18)$$

where again K' is the activity coefficient term. The complementarity of (4.18) and (4.10) is evident in this case as well.

Qualitative arguments that invoke the Donnan potential can be of help in explaining equilibria involving ions of different charge. A hydronium form resin in a dilute solution of hydrochloric acid will give rise to a large potential difference between the phases. A double charged ion such as calcium when placed in the solution will therefore be more strongly attracted to the resin than will be a singly charged ion such as sodium.

How is this preference affected by solution concentration? As the concentration of the solution is increased the potential diminishes. This is so because the potential is determined by the difference in escaping tendencies of the hydronium ion in the two phases and increasing the solution concentration reduces this difference, and theoretically eliminates it altogether when the hydronium ion ''pressure'' from the solution phase equals that from the resin phase. Along with this disappears the electrostatically based preference for the multiply charged ion. So not only does increasing concentration exert a mass action effect, which we would expect to be proportional to concentration, but the diminishing Donnan potential represents additional attenuation of the preference for the multiply charged ion. This extra influence makes the exponent on the concentration term intuitively more palatable.

Bonhoeffer[6] coined the term ''electroselectivity'' for this phenomenon. It is a physical effect that would persist even if ions behaved ideally as simple point charges. In that case one would be unable to separate sodium from potassium or calcium from barium but it would still be possible to separate sodium from calcium, for instance.

The term ''electroselectivity'' tends to suggest that this phenomenon can occur only with charged species, but Helfferich has pointed out that the selectivity arises from the stoichiometry of the exchange and not particularly because we are dealing with ions.[7]

4.2.4. A GENERAL EXPRESSION FOR K_D

The exchange of ions between an ion exchanger and a solution may be represented by the following general expression:

$$yR_xE^x + xS^y \rightleftharpoons xR_yS^y + yE^x \qquad (4.19)$$

Charges on S and E have been omitted so that the expression may be applied to either anion or cation exchange; R denotes the resin functionality and it is assumed to be singly charged.

Applying the mass action law to the equilibrium gives

$$[S^y]_R^x/[S^y]_S^x = K_E^S[E^x]_R^y/[E^x]_S^y \qquad (4.20)$$

where K_E^S is the equilibrium coefficient. Again, invoking the condition that in IC one ion, the eluent ion E^x, is abundant while the sample ion S^y is at comparatively much lower concentrations means that $[E^x]_R$ tends to C_R, the capacity of the resin.

Introducing this approximation gives

$$([S^y]_R/[S^y]_S)^x = K_E^S C_R^y/[E^x]_S^y \tag{4.21}$$

or K_D for

$$S^y = K_E^S C_R^{y/x}/[E^x]_S^{y/x} \tag{4.22}$$

If the constant term $K_E^S C_R^{y/x}$ is denoted simply by B, then $K_D = B/[E^x]_S^{y/x}$ or

$$\log K_D = -(y/x)\log[E^x]_S + \log B \tag{4.23}$$

This expression is commonly used and cited in IC work and its validity has been demonstrated for many systems.[8,9] For example, Small and Miller[9] showed that it applied to the elution of several inorganic ions by mono- and multivalent ions such as iodide, phthalate, sulfobenzoate, and trimesate (Figure 4.4). The agreement between the theoretical and experimentally derived slopes was in most instances found to be excellent (Table 4.3).

The expression, when it applies, has important practical implications; knowing the charges of the sample (y) and eluent ion (x) and making a determination of k-prime for an ion at but a single concentration it is then possible to

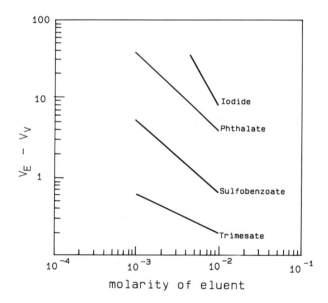

FIG. 4.4. The retention of sulfate ion as a function of eluent concentration for several eluents. The ion exchanger was a surface agglomerated anion exchanger. (From Ref. 9 with permission.)

TABLE 4.3. Comparison of y/x (Theory) and y/x (Observed)[a,b]

		y/x	
E^{x-}	S^{y-}	Theory	Observed
I^-	Br^-	1	1.00
	SO_4^{2-}	2	2.00
Trimesate^{3-}	Br^-	1/3	0.21
	SO_4^{2-}, $S_2O_3^{2-}$	2/3	0.60
	TPP^{y-}	?	1.9
Phthalate^{2-}	Br^-	1/2	0.47
	SO_4^{2-}	1	0.98

[a]Reference 9.
[b]Note: TPP^{y-} denotes tripolyphosphate ion.

construct plots such as Figure 4.4 and from them to broadly predict elution behavior.

Complications arise when more than one ion performs an eluting role and especially where the relative abundance of the two ions can change with concentration or pH. A good example is the carbonate–bicarbonate–hydroxide eluent system of anion IC, where elution behavior is not as predictable, at least not in a quantitative way. But for other important eluents such as hydronium for cations and hydroxide for anions the expression has great utility.

Whether or not the general expression is precisely applicable to a particular system is less important than one of its underlying concepts, namely, that the charge on eluent and sample ions has a profound influence on ion exchange elution behavior. The general intractability of eluites of high charge and the particular efficacy of eluent ions that are multicharged are two of the more important consequences.

As a rule, increasing the ionic strength of the eluent will hasten elution unless complexing is present, in which case the reverse can happen (see Section 4.2.7a). Thus, increasing the concentration of the eluent is one means, and an effective one, for bringing elution times into a practical range. However, there are many practical reasons for placing an upper limit on the concentration of the eluent: economy in the use of expensive reagents, the hazardous or corrosive nature of the eluent, disposal of waste, or a variety of inconveniences or problems attendant on using concentrated solutions.

Sensitivity of detection can be the most critical factor in limiting the level of eluent that may be used, and this is especially true when the eluent possesses the property to which the detector is sensitive. For a fuller discussion of this topic the reader is referred to the chapters on detection.

4.2.5. K_D and k' for Low-Capacity Resins

The capacity of an ion exchanger is a key factor in determining ion retention. Providing resins of varied capacity has been the basis for some of the major advances in modern IC. Before discussing the effects of capacity on chromatographic behavior of ion exchangers it is necessary to modify some of the relationships that have been developed for the fully functionalized or high-capacity materials.

For any ion exchanger the k-prime is by definition the ratio of the amount of material in the exchanger phase to the amount of material in the solution or mobile phase. The k-prime is related in turn to the distribution coefficient K_D as follows (see Section 2.3.3):

$$k' = [A]_R V_R/[A]_S V_S = K_D V_R/V_S \qquad (4.24)$$

where $[A]_R$ and $[A]_S$ are the concentrations of the distributing ion in the resin and solution phases, respectively, and V_R and V_S refer to the volumes of the resin and mobile phases.

For a column packed with spherical particles where the whole particle carries functionality, the active part of the stationary phase occupies roughly 65% of the column volume so the ratio V_R/V_S is approximately 1.86. For a pellicular exchanger where only a portion of the sphere is involved, the ratio is much less than this. To calculate k-prime values for pellicular materials it is necessary that we know, or at least have a reasonable estimate of, the volume of the pellicle. I know of no case where the volume of the pellicle has been measured, but it may be estimated from the capacities of the low- and high-capacity materials in the following simple manner. Let us assume that the measured (assumed) capacity of a pellicular material is denoted C_{pell} (milliequivalents per milliliter of resin) and that the specific capacity of a fully functionalized particle with properties identical to the pellicle is denoted C_R; then the volume fraction of the particle occupied by the pellicle is simply C_{pell}/C_R. In a column packed with pellicular particles, the volume of column occupied by *active* ion exchanger phase (V_R) is therefore C_{pell}/C_R times the volume occupied by the pellicular resin, that is

$$0.65V_{col}(C_{pell}/C_R) \qquad (4.25)$$

where V_{col} is the volume of the column.

The volume of the mobile phase V_s is of course the same as it is for a fully functionalized material, that is, $0.35V_{col}$. So V_R/V_S for a pellicular resin is

$$0.65V_{col}(C_{pell}/C_R)/0.35V_{col} \qquad (4.26)$$

or

$$1.86(C_{pell}/C_R) \qquad (4.27)$$

k-prime for a pellicular resin is therefore given by the expression

$$k' = 1.86(C_{\text{pell}}/C_R)K_D \qquad (4.28)$$

where K_D is the distribution coefficient calculated for a fully functionalized material (Section 4.2.2).

Equation (4.28) assumes that the distribution properties of the ion exchange active layer are independent of the fraction of the total resin volume occupied by the layer. In some instances this assumption is approximately valid, while in other cases it is not; it depends on the manner in which the surface layer is produced.

In the case of resins prepared by the surface agglomeration method (see Chapter 3) capacity is controlled by altering the following factors:

1. The degree of surface coverage of the substrate resin by colloidal ion exchanger.
2. The diameter of the substrate bead.
3. The diameter of the colloidal resin attached to the surface.

The distribution properties of the surface layer will be the same for all resins produced from the same colloidal component—which is usually the case when factors 1 and 2 are used to control the capacity of the agglomerated material. However, when the diameter of the colloid is changed, its distribution properties, and therefore K_D, can also change, although on strictly physical grounds we would not expect them to. Why should this be so? Colloidal resins of different diameter are usually prepared in different synthetic runs, which can lead to resins of different ion exchange properties even when the objective is to produce colloids that differ only in size. Undetermined factors in the synthesis are responsible for this.

Modification of a thin surface layer of an inert polymer particle is a widely used means for producing resins of different capacity. This too can lead to surface layers of different distribution properties even when the same substrate is used. The surface sulfonation of styrene divinylbenzene copolymers of either the gel or macroporous type to yield low-capacity cation exchangers is a well-known route to resins important to IC. Sulfonation tends to take place with a sharp boundary between the essentially fully sulfonated shell and the unmodified core. However, the boundary between reacted and unreacted core must necessarily be somewhat diffuse, and while most of the sites are in the highly swollen shell there are some sites buried in water-poor, hydrophobic regions near the boundary. These sites can be expected to express a selectivity quite different from those in more water swollen regions. The lower the degree of sulfonation the greater is the fraction of sites in the boundary region, so one would expect a variation not only in capacity but also in selectivity as the degree of sulfonation is changed.

In a similar connection, surface sulfonated cation exchangers display a selectivity that is not compatible with their cross-linkage; they appear to act like a resin of higher cross-linking. This may be due to a pseudo-cross-linking imparted by the resin core which, because it is unswollen itself, prevents the sulfonated layer from swelling to its fullest extent. One would expect this restriction to be more pronounced the lower the degree of surface sulfonation.

In summary, there are a variety of reasons for assuming that it is not just the capacity that is altered in these surface modification reactions. So in applying equation (4.28) it should be kept in mind that K_D may be varying as well as C_{pell}.

Regardless of these complications, it is generally true that in ion exchangers of this type the lower the capacity the lower th k-prime of an ion. Lowering resin capacity brings multiple benefits to IC: it lowers elution times, and it permits lowering of eluent concentrations, which in turn can have directly beneficial effects on the sensitivity of detection (Section 7.4.2). It also tends to improve mass transfer rates because of the reduced diffusional distance in the resin phase. About the only detrimental effect of lowering capacity is the increased tendency for the resin to become overloaded with sample.

A simple chromatographic exercise will show the effect of capacity on retention. We will consider the elution of a small amount of bromide ion from a column of anion exchange resin by a solution containing hydroxide ions— sodium hydroxide, for example. A column volume of 1 ml will be assumed, and if the resin is of medium cross-linkage and fully functionalized, the total capacity will be approximately 2 meq. It is valid for the main purpose of this exercise to use the selectivity coefficients of Table 4.2.

The selectivity coefficient for bromide–hydroxide exchange may be calculated from the values for chloride–hydroxide and chloride–bromide values:

$$K^{Br}_{OH} = K^{Br}_{Cl} K^{Cl}_{OH} = K^{Br}_{Cl} / K^{OH}_{Cl} = 2.8/0.09 = 31 \qquad (4.29)$$

The distribution for bromide ion in the system is given by

$$K_D = K^{Br}_{OH} C_R / [OH^-]_S \qquad (4.30)$$

If the concentration of the sodium hydroxide eluent is assumed to be 0.1 M then the distribution coefficient for the bromide eluite ion calculates to be

$$K_D = (31)(2)/0.1 = 620 \qquad (4.31)$$

and the k-prime in turn to be 1116, which is a very large value indeed, much too large for practical IC. For a 1 ml column of resin and assuming a flowrate of 2 ml/min, the elution time of the bromide ion would be over 3 h. Should the sample contain ions with a greater affinity for the resin than bromide, the elution of the total sample could take much longer than this. By reducing the resin's capacity a hundredfold the elution time of bromide is reduced proportionately to about 2 min, a much more acceptable time. This demonstrates one of the main arguments in favor of low-capacity resins. Changing resin capacity is only one part of designing chromatographic conditions, but it is a vital part of the total strategy of method development in modern IC.

4.2.6. Effects Attributable to the pH of the External Phase

The partitioning of an ion between a resin and a solution phase can be profoundly modified if that ion undergoes hydrolytic reactions, for example,

$$H_2CO_3 \rightleftharpoons HCO_3^- + H^+ \rightleftharpoons CO_3^{2-} + H^+$$
$$H_2PO_4^- \rightleftharpoons HPO_4^{2-} + H^+ \rightleftharpoons PO_4^{3-} + 2H^+$$
$$NH_3^+ \, RCOOH \rightleftharpoons NH_3^+ \, RCOO^- + H^+ \rightleftharpoons NH_2 \, RCOO^- + 2H^+$$

Examples abound where a change in the pH of the aqueous environment may be used to either increase or diminish the sorption of an ion onto an ion exchanger. The charge of an eluite ion may be altered to either retard or promote its elution. Similarly, the potency of an eluent may be manipulated simply by changing the pH; raising the pH of carbonate–bicarbonate systems increases the concentration of the divalent species, leading to greater displacing power; the conversion of biphthalate to phthalate has a similar effect on phthalate eluents.

Expressions can be developed that relate distribution coefficients, ion exchange equilibria, and acid–base dissociation equilibria, but they tend to be involved and cumbersome to use. Furthermore, reliable ion exchange equilibria data for related pairs such as $HCO_3^- - CO_3^{2-}$ are not readily available. Logarithmic diagrams (see Appendix A) combined with a qualitative appreciation of ion charge effects can be of considerable help in predicting or explaining pH effects.

In some ion exchange reactions it is not only the pH of the external phase that is important but the pH of the resin phase as well. How does one determine the pH of the resin phase? It cannot be measured, but it is possible to estimate it, as the following example will demonstrate. Consider a sulfonic acid-type cation exchanger in contact with a solution of sodium and hydronium ions, $10^{-2} \, M$ and $10^{-4} \, M$ in each, respectively. The pH of the solution will be approximately 4. We will assume that the capacity of the resin is 3 meq/ml, and for the purposes of this calculation it is permissible to ignore selectivity and assume that the resin shows no preference for sodium over hydronium. Therefore, at equilibrium the ratio of sodium and hydronium inside the resin will be the same as outside, that is, 100 : 1. Since the resin will be mainly in the sodium form, the concentrations of sodium and hydronium ions in the resin phase will be $3 \, M$ and $3.10^{-2} \, M$, respectively. Thus the pH *in the resin phase* will be less than 2, more than two units lower than the solution phase with which it is in contact and in equilibrium. A similar argument for the other end of the pH scale will show how the basicity of an anion exchange resin phase can be much greater than the solution phase. These differences between the resin and solution pH have important consequences for the distribution of species that undergo protolytic reactions. The amino acids are an important example of such behavior; their acid–base chemistry and ion exchange behavior are treated in more detail in Chapter 9.

The question arises from time to time if ion chromatography can be used to separate closely related species such as HPO_4^{2-} and PO_4^{3-}. It cannot. It is true that the HPO_4^{2-} species, being of lower charge, will tend to elute more rapidly than the triply charged PO_4^{3-} ion. There would be an inclination therefore for the upstream side of the band to become enriched in PO_4^{3-} and the downstream side in HPO_4^{2-} were it not for the instantaneous reestablishment of the hydrolytic equilibrium to maintain a constant ratio of phosphate species in the phosphate zone. Thus the two species cannot possibly disengage in the column—any tendency to do so is corrected by the dissociation (association) reactions.

4.2.7. Complexation and Ion Exchange Equilibria

Ion exchange equilibria are strongly dependent on interactions between the counterions and other species in the solution phase. Specific interactions involving hydronium ion with hydroxide ion and other bases, and their importance for IC have already been noted. Complexing of metal ions by anionic and neutral ligands comprise another very important class of reaction.

Many metal ions will form complexes with simple inorganic ions such as F^-, Cl^-, SO_4^{2-}, CO_3^{2-}, CNS^-, and CN^-, and with multidentate organic ligands such as polyamines and polyamino carboxylic acids like EDTA and NTA. Factors that are important to the ion exchange behavior of such complexes are the charge of the complex, its stability, and the affinity of the complex for the exchanger.[10]

Of these, perhaps the most important factor influencing the ion exchange of complexed metal is the charge of the complex. Complexation of a metal ion with an anionic ligand can change it all the way from being cationic through neutral to anionic. Changes of this sort can be effected gradually, thereby diminishing the affinity of the metal species for a cation exchanger while increasing its affinity for an anion exchanger.

The stability of the complex will determine the level of ligand that must be supplied to the external phase to promote significant complexing. Chloride ion forms rather weak complexes with most metal ions so it requires rather high concentrations of ligand to form significant levels of complex. Present day IC practice tends to shun the use of concentrated eluents particularly of corrosive reagents such as hydrochloric acid, a popular complexing medium. But some metals, notably gold and mercury, form chloro-complexes quite readily and are more amenable to the standard IC environments. Cyanide complexes of silver, gold, copper, cobalt, etc. are quite stable and are likewise amenable to modern IC practice. Multidentate chelating ligands such as EDTA can form extremely strong complexes with alkaline earth, transition metal, and lanthanide ions and have an important place in IC methodology.

If metals form anionic metal complexes then they can be taken up by an anion exchanger and their distribution will be affected by the usual factors: the charge of the complex, the concentration of competing anions, and the affinity of the complex for the exchanger relative to competing anions in the system. A weak tendency to complex can sometimes be overshadowed by a very high affinity for the resin phase; the extraction of chloroferrate complexes into anion exchangers appears to be a good example of this. So high is the affinity of these complexes for an organic resin matrix that chloroferrate complexes will even extract into a cation exchanger in concentrated hydrochloric acid when extraction of cationic iron is unlikely. Iron has even been observed to adsorb from concentrated hydrochloric acid onto a macroporous styrene divinylbenzene polymer, which is nominally considered to be neutral and "inert."[11]

Complexing systems can be divided into two main groups: those involving the common inorganic ligands and those involving organic chelating agents and other multidentate ligands.

4.2.7a. Inorganic Ligands

Kraus and his co-workers in the early 1950s were the first to make a comprehensive study of inorganic complexes in ion exchange. They examined and documented the distribution of a large number of metal ions between hydrochloric acid and an anion exchanger, Dowex 1. They reported their data mainly in the form of plots of distribution coefficients versus acid concentration[12]; two examples from their work are shown in Figure 4.5. A number of other groups also studied inorganic complexes and devised many analytical schemes for the separation of metals based on this unique ion exchange–complex chemistry.[13–15] But apart from the cases already mentioned above for chloro- and cyanocomplexes, most of this classic work has yet to be adopted into contemporary IC art. This may be due partly to the impracticality of using concentrated, sometimes toxic, sometimes corrosive reagents and partly to the lack of suitable detection schemes, although many complexes have usefully large molar extinction coefficients.[16]

Solvent effects are important in these complexing, ion exchange reactions. Complexes that form with difficulty in water may form readily in water solvent mixtures and the distribution of complexes between resin and solution can also be profoundly affected by solvent composition. This is another dimension of ion exchange behavior that has yet to be examined to any great extent in the context of modern IC. In this connection the book by Marcus and Kertes[17] is recommended for its depth and for its linking of the related areas of ion exchange and solvent extraction of metal complexes.

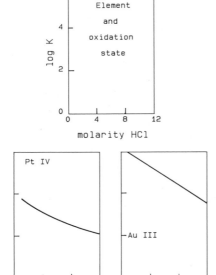

FIG. 4.5. Extraction of metal ions from concentrated hydrochloric acid solutions onto an anion exchanger. Distribution coefficients of Pt(IV) and Au(III) between an anion exchanger and hydrochloric acid as a function of HCl concentration. (From Ref. 12.)

Besides its analytical implications, the ion exchange of metals in concentrated solutions provides insights into the causes of ion exchange selectivity. The work of Diamond and co-workers is noteworthy and is discussed more fully elsewhere (Section 4.2.8).

4.2.7b. Organic Ligands

Complexation of metal ions by organic ligands is widely employed to facilitate difficult metal ion separations by ion chromatography. Ligands such as oxalate, tartarate, citrate, ethylenediamine tetraacetate (EDTA), and pyridine dicarboxylate (PDCA) have been used in conjunction with both cation and anion exchangers to separate a wide variety of metal ions. There is much that is understood about these systems that enables rational design of separation methods but much remains to be learned and separation schemes must often be developed in an empirical fashion. This is particularly so for the separation of metal complexes on anion exchangers.

Understanding the chromatographic behavior in these systems requires an understanding of the interplay between ion exchange and complexation reactions. To simplify the development of principles, the discussion will be restricted, for the time being, to a limited set of metal ions and ligands. We will consider the equilibrium distribution and chromatography of Co, Cu, Ni, and Zn on both cation and anion exchangers with oxalate ion as the complexing ligand. The ion exchangers are of the strong acid or strong base type.

First we will consider the distribution of a metal ion M^{2+} between a cation exchanger and a solution containing sodium ions as the other competing counterion. The distribution is described by

$$K_D = [M^{2+}]_R/[M^{2+}]_S = K_{Na}^M [Na^+]_R^2/[Na^+]_S^2 \qquad (4.32)$$

where K_{Na}^M is the selectivity coefficient for the exchange reaction

$$2R\text{–}SO_3^- Na^+ + M^{2+} \rightleftharpoons (R\text{–}SO_3^-)_2 M + 2Na^+ \qquad (4.33)$$

Making the usual approximations when M^{2+} is present at trace levels, that is, $[Na^+]_R = C_R$, we have

$$K_D = K_{Na}^M C_R^2/[Na^+]_S^2 \qquad (4.34)$$

Table 4.1 indicates that a sulfonic acid resin shows very little spread in selectivity between these four subject ions, and a simple elution with salt or acid would be expected to give an inadequate separation.

Adding a complexing agent such as oxalate to the solution phase shifts the locus of selectivity to the external phase. Complexing of metal ion by oxalate can be considered as a two-step reaction:

$$M^{2+} + Ox^{2-} \rightleftharpoons MOx \qquad (4.35)$$

$$MOx + Ox^{2-} \rightleftharpoons MOx_2^{2-} \qquad (4.36)$$

Each step can be described by a mass action equilibrium:

$$K_1 = [MOx]/[M^{2+}][Ox^{2-}] \qquad (4.37)$$

$$K_2 = [MOx_2^{2-}]/[MOx][Ox^{2-}] \qquad (4.38)$$

If oxalate is in considerable excess over the metal ion concentration, as it will be in the typical chromatographic experiment, then, given the magnitude of K_1 and K_2, the species MOx_2^{2-} will be favored and the equilibrium may be written

$$M^{2+} + 2Ox^{2-} \rightleftharpoons MOx_2^{2-} \qquad (4.39)$$

for which the mass action expression is

$$B = K_1 K_2 = [MOx_2^{2-}]/[M^{2+}][Ox^{2-}]^2 \qquad (4.40)$$

where B is the complexation constant for reaction (4.39). Values of B for some transition metal ions are provided in Table 4.4.

As metal is complexed it loses its cationic character while at the same time it develops anionic character in the form of the doubly charged metal oxalate com-

TABLE 4.4. Stability Constants for
Metal Oxalate Complexes[a]

M^{2+}	$\log B^b$
Co	5.6
Cu	10.27
Ni	6.6
Zn	6.0

[a]Reference 18.
[b]B as defined in equation (4.40).

plex. Complexing can therefore be expected to have the effect of lowering the distribution of metal ion into a cation exchanger while promoting its distribution into an anion exchanger. Consider the cation exchange case first.

The distribution coefficient of the metal in the system is the ratio of the concentration of metal species in the resin to their concentration in the external phase. The metal species in the resin phase is the uncomplexed cation, M^{2+}, and its concentration is $[M^{2+}]_R$. The solution phase contains the species $(MOx_2)^{2-}$ in addition to M^{2+} so the concentration of metal in the external solution is $[M^{2+}]_S + [MOx_2^{2-}]_S$. From (4.40) we have

$$[MOx_2^{2-}] = B[M^{2+}]_S[Ox^{2-}]^2 \qquad (4.41)$$

so that K_D for the distribution of metal in the complexing system is

$$K_D = [M^{2+}]_R/([M^{2+}]_S + B[M^{2+}]_S[Ox^{2-}]^2) \qquad (4.42)$$

Combining (4.41) and (4.42) gives

$$K_D = [M^{2+}]_S K_{Na}^M C_R^2/[Na^+]_S^2\{[M^{2+}]_S + B[M^{2+}]_S[Ox^{2-}]_S^2\} \qquad (4.43)$$

$$= K_{Na}^M C_R^2/[Na^+]_S^2\{1 + B[Ox^{2-}]_S^2\} \qquad (4.44)$$

From this we can draw a number of conclusions:

1. With no oxalate present the term $1 + B[Ox^{2-}]_S^2 = 1$ and the expression is identical to (4.34)—that is, the exchange is totally controlled by the cation exchange equilibrium, as is to be expected.

2. As oxalate is added, the distribution favors metal in the external phase. Raising the pH of the solution will have a similar effect since that increases the concentration of Ox^{2-}.

3. For a fixed level of oxalate, the spread in stability constants will be reflected in a corresponding spread in K_D values for the various ions. Since there is a much bigger spread in stability constants than there is in selectivity coefficients, we would expect more successful separation of metal ions with oxalate present than in its absence.

Figure 4.6 is a plot of calculated distribution coefficients as a function of oxalate concentration. In these calculations the capacity of the resin is assumed to be 3 meq/ml and the concentration of sodium in solution 0.2 M. The marked spread in distribution coefficients suggests that a good separation is possible.

Figure 4.7 is the result of an actual separation carried out on a surface sulfonated resin using an oxalate eluent; the completeness of the separation and the order of elution indicate a harmony between theory and practice. This is the case with most separations using cation exchangers in conjunction with complexing ligands.

In presence of anionic ligands metal ions develop anionic character and will be picked up by and separate on anion exchange resins. The chromatography will again depend on the charge of the complex, the stability of the complex, and the affinity of the complex for the anion exchanger.

The charge of the complex is probably the single most important factor controlling chromatographic behavior in these systems. The charge in turn depends on

1. The charge of the central metal ion.
2. The charge on the ligand.
3. The number of attached ligands.

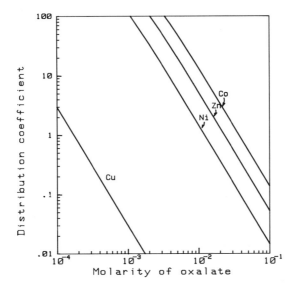

FIG. 4.6. Calculated distribution coefficients for Co^{2+}, Cu^{2+}, Ni^{2+}, and Zn^{2+} in the system: Sulfonic acid type resin–Na^+ (0.2 M)–Na^+ oxalate. The concentration of sodium ion was fixed at 0.2 M while the concentration of oxalate was varied. The capacity of the resin was assumed to be 3 meq/ml.

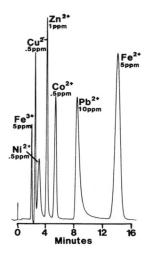

FIG. 4.7. The separation of transition metals on a surface sulfonated cation exchanger using oxalate as a complexing agent in the mobile phase. (From Ref. 20 with permission.)

Knowing or determining the charges on the metal ion and its ligand is usually a straightforward matter. The number of attached ligands is dictated by the coordination chemistry of the metal—ligand system; it will depend mainly on the coordination requirements of the metal ion and the donor characteristics of the ligand.

In oxalate media, ions of transition metals such as Co, Cu, and Ni tend to coordinate two moles of ligand, so the net charge on the complex is minus two:

$$M^{2+} + 2Ox^{2-} \rightleftharpoons (M^{2+}Ox_2^{2-})^{2-} \tag{4.45}$$

Trivalent iron on the other hand will bind three moles of oxalate giving a net charge of minus three:

$$Fe^{3+} + 3Ox^{2-} \rightleftharpoons (Fe^{3+}Ox_3^{2-})^{3-} \tag{4.46}$$

Lanthanide ions will also bind three moles of oxalate but can bind as many as four, giving complexes of minus three and minus five charge: $(LnOx_3)^{3-}$ and $(LnOx_4)^{5-}$. Highly charged species such as these can be expected to have a high affinity for anion exchangers. Furthermore their elution behavior can be expected to be very sensitive to the charge and concentration of the displacing ion (see Sections 4.2.3 and 4.2.4).

Multidentate ligands may contain sufficient donor sites that the coordination needs of the metal ion may be satisfied with but relatively few ligands. Polyamino-carboxylate ligands such as EDTA are good examples; they contain several nitrogen and oxygen donor groups and most metal ions bind but one molecule of ligand. Since ligands such as this contain uncharged as well as anionic sites a variety of possibilities exist for the charge of the complex. The complexing of cupric and ferric ions with a variety of such chelants illustrates the various possibilities:

$$Cu^{2+} + EDTA^{4-} \rightarrow (Cu\ EDTA)^{2-}$$
$$Fe^{3+} + EDTA^{4-} \rightarrow (Fe\ EDTA)^{-}$$
$$Cu^{2+} + EDDA^{2-} \rightarrow (Cu\ EDDA)^{0}$$
$$Cu^{2+} + NTA^{3-} \rightarrow (Cu\ NTA)^{-}$$

(EDDA is ethylene diamine diacetate, NTA is nitrilo triacetate.) Not surprisingly this leads to a variety of chromatographic behavior.

If more than one ligand or chelating species is present, then, as well as forming complexes with the separate ligands, it is also possible for the metal ion to form complexes that involve more than one ligand in the same molecule. Evidence for this sort of behavior was observed in the chromatography of copper–NTA complexes.[19] When Cu–EDTA and Cu–NTA are eluted from an anion exchanger using chloride as displacing ion, the NTA complex elutes ahead of the EDTA complex as would be expected from their charges—minus one for the former, minus two for the latter. When carbonate is added to the system the elution order is reversed, the NTA complex being considerably more resistant to elution than the EDTA complex. To explain the reversal, it is argued that NTA cannot completely supply the coordination needs of the tetra-coordinate cupric ion and that the remaining position is occupied by carbonate when it is present:

$$(Cu\ NTA \cdot H_2O)^- + CO_3^{2-} \rightleftharpoons (Cu\ NTA \cdot CO_3^{2-})^{3-} + H_2O$$

The substitution of water by carbonate in the coordination sphere would yield an ion of minus three charge, which could explain its eluting after the Cu–EDTA^{2-} complex, which is coordinatively satisfied by one EDTA and does not pick up further ligand.

In summary, the charge of the metal ion, the charge of the ligand, and the coordination number of the metal ion play vital parts in determining one of the most dominant factors in ion chromatography, namely, the charge of the eluite ion.

For metal ions of the same charge and similar chemistry an increase in complex stability implies an increased tendency to form anionic complexes if the ligand is anionic. This implies in turn an increased affinity for anion exchangers all other things being equal. The data to some extent bear this out. Figure 4.8 shows the separation of oxalate complexes of a number of transition metal ions on a pellicular (surface agglomerated) anion exchanger. The order of elution of the three cations Co, Ni, and Zn is what one would expect from the relative stabilities of their oxalate complexes; the ion with the greatest tendency to complex elutes last. The position of copper, on the other hand, indicates that factors other than stability are involved; since copper forms the most stable complex of all the ions in the set it should elute after all the others. Instead it

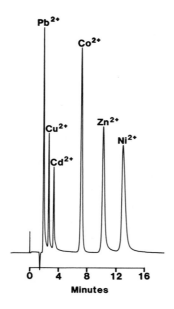

FIG. 4.8. Separation of oxalate complexes of transition metal ions on an anion exchange resin. (From Ref. 20 with permission.)

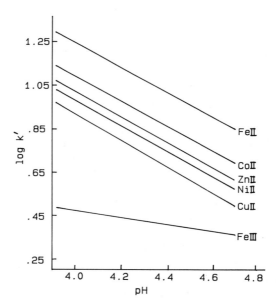

FIG. 4.9. Retention (*k*-prime) of transition metal PDCA complexes on an anion exchange resin. (From Ref. 21 with permission.)

elutes earlier than most. The chromatography of pyridine dicarboxylate complexes is similarly perplexing. The order of elution of the metal complexes (Figure 4.9) is the opposite to the order predicted by complex stability alone (Table 4.5).

It may be speculated that the metal complexes have quite different affinities for the anion exchangers and that it is this factor that overrides the complexing effects. But until a definitive study has been made, much of the anion exchange chromatography of metal complexes will remain unpredictable and method development will of necessity tend to be empirical.

4.2.7c. Complexing as a Means of Separating Ligands

Complexing reactions, as well as aiding metal ion separation, can also be used to separate ligands. The separation of the copper complexes of EDTA and NTA is an example of this. Complexing can add a number of benefits to this kind of separation: it can subdue the charge of certain highly charged chelants making them more elutable, and coupling of the chelant with ions such as copper or iron can render them more amenable to detection, e.g., by UV photometry (see the separation of polyamines in Chapter 9).

4.2.7d. Ligand Exchange

Ligand exchange is a means of separation that predates modern IC. It was first proposed by Helfferich[22] and has been more recently reviewed by Davan-

TABLE 4.5. Stability Constants for Metal–PDCA[a] Complexes[b]

Metal ion	Equilibrium	log K
Fe^{2+}	$[ML]/[M] \cdot [L]$	5.71
	$[ML_2]/[M] \cdot [L]^2$	10.36
Co^{2+}	$[ML]/[M] \cdot [L]$	6.65
	$[ML_2]/[M] \cdot [L]^2$	12.70
Zn^{2+}	$[ML]/[M] \cdot [L]$	6.35
	$[ML_2]/[M] \cdot [L]^2$	11.88
Ni^{2+}	$[ML]/[M] \cdot [L]$	6.95
	$[ML_2]/[M] \cdot [L]^2$	13.5
Cu^{2+}	$[ML]/[M] \cdot [L]$	9.14
	$[ML_2]/[M] \cdot [L]^2$	16.52

[a]PDCA, pyridine-2,6-dicarboxylate; L denotes PDCA ligand.
[b]Reference 18.

kov and Semechkin.[23] It may be used in either a preparative or an analytical way; we will consider it in the latter context.

The basis of ligand exchange is the selective sorption and displacement of ligands on ion exchange resins bearing complexing cations such as Cu^{2+}, Ni^{2+}, Hg^{2+}, or Ag^+. The ligands may be anionic or they may be neutral such as amines, ammonia, carboxylic acids, and amino carboxylic acids. In a typical ligand exchange separation, the metal ion form of the resin is first equilibrated with a solution containing a ligand, the displacing ligand E_L, that incorporates into the coordination sphere of the ion on the resin. Sorption of E_L will generally involve displacement of some other species from the coordination shell of the metal ion. This might be water from the hydration shell of the ion if a simple cation exchange resin is used, or it might be a ligand supplied by the functional group of the resin if some sort of complexing resin is used.

In the second step, sample containing a mixture of potential ligands is loaded to the column, at which point the ligands are incorporated into the coordination shell of the bound ion. Elution then begins with the ligand from the eluent selectively displacing the sample ligands from the metal ions. Eventually sample ligands elute from the column in ascending order of their affinity for the bound metal ion.

Ideally, metal ion should not be displaced from the resin during use. Theoretically, no metal ion will be lost when the eluent and sample contain only neutral species, but, whereas it may be possible to devise a neutral eluent, it is common for sample to contain electrolyte whose ions will tend to remove metal from the resin. This can be counteracted to a great extent by using complexing resins that bind the metal ion more securely but at the expense of reduced ligand binding capacity since some of the coordination sites on the metal ion are now occupied by a ligand(s) from the resin.

Ligand exchange has been applied to the separation of a wide range of organic species as well as some inorganic,[23] but as Davankov and Semechkin point out, its application in automated systems is hampered by detection problems. The ligand exchange chromatography of amines is a good example of this. The common eluent for this separation is ammonia, which presents a formidable problem for the sensitive detection of amines against this ammonia background. Some ligand exchange reactions can be relatively sluggish, and this can lead to rather poor efficiencies from ligand exchange columns.

Despite these problems, ligand exchange has many potential uses in chromatography and in analysis generally. It can be used to separate optical isomers, to concentrate complex forming species from extremely dilute solutions of the same, and to sorb biopolymers. It is the basis for the extraction of species from organic media and it has even been used to remove species from the gas phase.[23,24] Ligand exchange should be regularly examined as a possible solution to analysis problems, especially in light of modern developments in resins and in detection techniques.

4.2.8. Theories on the Causes of Ion Exchange Selectivity

The selectivity of ion exchangers provokes interest in its causes. From a fundamental point of view there is the challenge of finding a unifying theory that accommodates the widely diverse behavior of exchangers of all types. On a more pragmatic level, a good understanding of the origins of selectivity should enable us to manipulate and control this most important of all ion exchange properties. At the present time there is no general theory of selectivity but there are a number of persuasive qualitative arguments that provide useful guidelines for the rational design of new ion exchangers. We will start with cation exchange of monovalent ions for that is the area that has been the most studied.

In sulfonated resins, the most common cation exchangers, the order of affinity usually observed is Li < Na < K < Rb < Cs with hydronium ion falling between sodium and lithium. This order is so familiar that it is often referred to as the *normal* order. An early theory by Gregor[25,26] affirmed that the size of the hydrated ion dictated the order of selectivity, with the resin preferring the ion of smaller hydrated diameter. Stress in the elastic network of the cross-linked resin was the key feature of this explanation. Thus the swelling stress in the resin was diminished and energy lowered by incorporating ions of smaller hydrated size. From ion mobilities, lithium was taken to be the largest hydrated ion and cesium the smallest, hence the observed selectivity order. The Gregor theory also accommodated the observation that as the resin was converted to the more favored ion form the selectivity for that form diminished (Figure 4.2). As the resin was converted to the more favored ion form it was argued that in this less stressed form it had less "need" to prefer that ion over the one of larger hydrated size. Where the Gregor theory and others fell short was their inability to provide a reasonable explanation for selectivity reversals. In sulfonated resins of higher cross-linking it was often observed that the normal order of selectivity could be reversed, especially when one of the ion pairs being studied was the hydronium ion. Thus in Figure 4.2 the highly cross-linked resin prefers sodium when it is mostly in the hydrogen form but hydrogen when mostly in the sodium form. Figure 4.10 shows a similar behavior for potassium–cesium selectivity; a "normal" order for low to moderate cross-linking but reversals for highly cross-linked versions. Bregman[27] and Reichenberg[3] also observed reversals in carboxylate resins that preferred lithium over sodium over potassium, although this order reverted to the more usual order for carboxylate resins that were only partially ionized.

Important researchers by Eisenmann on the electrochemical properties of a series of aluminosilicate glasses revealed a whole variety of selectivity sequences.[28] By altering the aluminum content, Eisenmann was able to change the order from

$$Cs > K > Na > Li$$

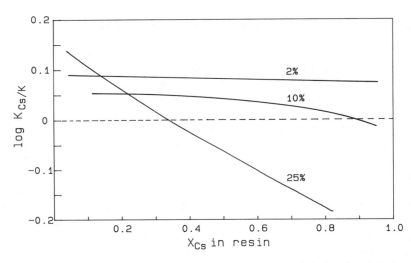

FIG. 4.10. Cesium–potassium selectivity on sulfonated cross-linked resins showing affinity reversals. (From Ref. 3 with permission of Marcel Dekker, Inc.)

through

$$K > Na > Li > Cs$$

to

$$Na > Li > K > Cs$$

From this work Eisenmann proposed a theory that is based on two main effects:

1. Electrostatic interaction between the functional group and the counterions.
2. Hydration energies of the competing counterions.

The functional group and counterion are treated as incompressible spheres with charge located at their center so that the electrostatic contribution to the energy was simply

$$e^2(r_A + r_1) \qquad (4.47)$$

where e is the electronic charge, r_A is the radius of the functional group, and r_1 is the radius of one of the pair of cations involved in the exchange. The close approach of counterion to functional group requires the loss or rearrangement of their hydration layers and this involves hydration energies as distinct from sizes of hydrated layers. Reichenberg[3] showed that the net change in free energy on removing ion M from solution and exchanging it for ion N on the functional group can be expressed as

$$\Delta G_{M/N} = [e^2/(r_A + r_N) - e^2/(r_A + r_M)] - (\Delta G_N - \Delta G_M) \quad (4.48)$$

where ΔG_M and ΔG_N are the standard free energies of hydration of the respective cations.

If the functional group is large (large r_A)–Eisenmann called this a fixed group of low field strength–the first term in square brackets on the right-hand side will be small regardless of the size of the cations and the second term will be the dominant one. Thus if ion M is potassium and N is sodium, since $\Delta G_N > \Delta G_M$, the total free energy change will be negative and the resin would prefer potassium. In this picture the sulfonate group is considered to be a large group of low field strength, which would explain the commonly observed order among the alkali metals. This also holds for the aluminate group in Eisenmann's glasses, which is considered to be low field strength. On the other hand if the functional group is small with a high field strength, such as carboxylate or silicate, then the first term dominates and it is the smaller ion that is preferred, leading, in the case of glasses of low aluminum content, to a complete reversal of the "normal" order. (Reichenberg points out that the "normal" order is nothing more than accidental.)

In the glasses, reversals are attributed to the dominant effect of the electrostatic term in moving from low field strength aluminate to higher field strength silicate; the cause of reversals in sulfonated and carboxylated resins is less apparent. Reichenberg argues that overlapping of the fields of the fixed group reinforces the electrostatic interaction and makes it the dominant factor. This overlapping is expected to be more pronounced in resins of higher capacity, in highly cross-linked resins, and in the more highly ionized forms of carboxylate resins.

The selectivity sequences for anion exchange vary widely. Here there is little correlation between hydrated ion size and affinity. The limiting conductance of the perchlorate ion, for example, is appreciably less than that of the chloride ion yet it is well known as having a much higher affinity for a quaternary ammonium-type anion exchanger (Table 4.2). Reichenberg has examined a number of factors that might be expected to dictate selectivity, such as hydrated size and polarizability but concludes that the electrostatic/hydration theory also adequately explains the order of anion selectivities.

Diamond and Whitney offer a somewhat different explanation for the causes of selectivity.[29] Their theory also stresses the importance of hydration but focuses on the differences between the resin and external phases with respect to water structure and water availability in the two phases. In their model the water of the external phase has the highly ordered structure of pure water that resists the intrusion of other species unless they have the ability to reorder or reorient the water. They see small ions with high hydration energies as having this ability while larger ions, or molecules, do not; in fact they can lead to "tightening up" of the water structure so that the external water phase is not a comfortable environment for such species and they tend to be rejected in favor of smaller species. On the other hand, in the resin phase the hydrocarbon matrix of the resin fragments the water into "pools and capillaries" and the water is less prone to reject large species.

Because the resin can be considered as a sort of concentrated solution, much of its water is used up in hydrating the ionic groups and there is less free water available. So an ion that requires more water of hydration will be more likely to satisfy its requirements in the external phase than in the resin. For simple ions their hydration needs can be correlated with ion size, fluoride ion having the strongest interaction with water and iodide the least. In the view of Diamond and Whitney the fluoride ion will be most easily accommodated in the external aqueous phase, and in competition with iodide, for example, the latter will be forced to the resin phase.

They also correlate anion affinity with base strength: the more strongly an anion picks up a proton the more strongly it bonds with water and in turn prefers the external aqueous phase. They cite the selectivity sequence of the anions trimethylacetate, methyl dichloroacetate, and trichloroaceate as supporting this. These ions have approximately the same size but decrease in basicity from trimethylacetate to trichloroacetate. It is the more basic that have the higher affinity for the external phase.

Diamond and Whitney also see parallels with the salting in of large molecules by large ions in aqueous solution. Concentrated solutions of quaternary ammonium salts dissolve appreciable amounts of "water-insoluble" molecules such as benzene.[30] The low water solubility of such nonpolar species is attributed to their inability to force cavities in the water structure. However, they can share cavities already formed by large-molecule ions and hence display significant solubility in the presence of such ions. Diamond and Whitney see the large functional groups of anion exchangers as having a similar effect of intruding into the internal water structure and forming hospitable sites for large ions rejected by the external water phase. They invoke the same argument to explain the selectivity order for the exchange of trialkyl ammonium cations:

$$NMe_4^+ < NEt_4^+ < NPr_4^+$$

On purely electrostatic grounds a cation exchange resin should prefer the smaller ion; the fact that opposite is the case supports the water structure based argument over the elctrostatic one.

The Diamond theory tends to emphasize the external phase as the seat of selectivity with the ion of weaker hydrating power forced into the resin phase as it were by default. For large "hydrophobic" ions the resin phase is seen as having a more direct role. Here hydrophobic interactions between the functional group and the counterion are important, and this is supported by cross-linking studies. As cross-linking is lowered the discrimination between small, highly hydrated ions diminishes, presumably because the difference between the resin and external phase diminish as the water content of the resin rises. But with bulky ions the selectivity remains high even with low cross-linked resins. In this case water enforced ion pairing is thought to be the major controlling factor.

Reversals in selectivity are also known in anion exchange. Schmitt and Pietrzyk,[31] studying anion exchange on alumina, reported retention orders of $F^- > Cl^- > Br^- > I^- > ClO_4^-$, which is just the opposite of that found in quaternary ammonium ion exchange resins. No explanation was offered for the observed order, but it suggests that the electrostatic term of the Eisenmann theory is dominant, as would be expected if the central aluminum ion is the seat of the charge—a high field strength site that favors small ions such as hydroxide and fluoride.

For inorganic ion exchangers such as this, many of the ion selectivity features can also be correlated with Pearson's concepts of "hard" and "soft" acids and bases.[32]

In a recent study Slingsby and Pohl report on the anion exchange selectivity of pellicular latex-based resins for several common inorganic ions.[33] They made a systematic examination of a number of resins where the quaternary ammonium functional group was varied but the degree of cross-linking and the polymeric backbone structure remained the same. Four amines were used to prepare the colloidal anion exchange resin particles: methyldiethanolamine (MDEA), dimethylethanolamine (DMEA), trimethylamine (TMA), and tri-ethylamine (TEA). The order of decreasing hydrophilicity of the resins was MDEA > DMEA > TMA > TEA. These workers were able to correlate a body of elution behavior with the hydrophilicity of the resin phase and the hydration enthalpies of the competing ions. Thus hydroxide ion was found to be an effective displacing ion on the more hydrophilic resins but much less so as the resin

TABLE 4.6. Retention Data[a] (k') for Anions on Resins of Different Hydrophobicity[b]

Resin	Analyte						
	F^-	Cl^-	Br^-	NO_3^-	ClO_3^-	SO_4^{2-}	PO_4^{3-}
MDEA	0.06	0.24	0.92	1.1	1.0	0.20	0.31
DMEA	0.14	1.1	4.5	5.0	4.9	3.0	6.7
TMA	0.30	4.4	19.2	22.5	21.2	51.4	>100
TEA	0.30	5.8	26.1	55.8	35.7	24.9	>100

[a]k' correlated for column capacity and normalized to the DMEA column. The eluent was 0.1 M sodium hydroxide.
[b]Reference 33.

TABLE 4.7. Retention Data[a] (k') for Anions on Latex-Based Resins[b]

Resin	Analyte						
	F^-	Cl^-	Br^-	NO_3^-	ClO_3^-	SO_4^{2-}	PO_4^{3-}
MDEA	0.13	0.38	0.86	0.74	0.40	1.3	6.5
DMEA	0.30	0.92	2.4	2.1	2.1	6.5	38.2
TMA	0.34	1.4	3.4	3.3	3.3	15.0	112
TEA	0.24	0.77	2.4	2.8	2.6	13.8	>150

[a] k' corrected for column capacity and normalized to the DMEA column. The eluent was 20 mM NaOH/5 mM p-cyanophenol.
[b] Reference 33.

hydrophilicity decreases. Remarkably, on the MDEA-based resin, of the ions examined, sulfate eluted earlier than all ions but fluoride (Table 4.6).

The more hydrophobic the resin the more it retained ions of lower hydration enthalpies—nitrate, bromide, and chlorate, for example. Likewise, the addition of a hydrophobic displacing ion, cyanophenate, to the eluent gave effective displacement of these ions (Table 4.7). Noteworthy also is the separation of nitrate and chlorate on the most hydrophobic resin whereas on the other resins these two ions essentially coelute. The greater retention of nitrate is also perplexing since it has a significantly higher enthalpy of hydration than does chlorate.[33] This is also the reverse of the order of elution in paired ion chromatography where chlorate is the more retained of the two (see Section 4.6). The authors attribute this ''anomalous'' retention of nitrate to specific interaction between the pi-electron clouds of nitrate and the aromatic rings of the polymer backbone. That cyanophenate causes a larger reduction in nitrate retention than it does in chlorate retention is also explained by this pi-overlap concept.

4.3. ION EXCHANGE KINETICS

The rate at which ions exchange between a resin and the external phase can be important in determining the efficiency of ion chromatography columns. Although not as intensively studied as equilibrium processes, the kinetics of ion exchange has been the subject of a great deal of theoretical and experimental effort. Important early work on ion exchange kinetics has been thoroughly summarized by Helfferich[6]; in this book we will consider some of the principal conclusions of this extensive literature.

As a general rule the rate of ion exchange processes is controlled by diffusion of the exchanging counterions. There has not been an authenticated instance of rate being controlled by a chemical reaction of the counterion with the functional group although Riviello has presented convincing evidence for chemical rate control in systems involving complexation reactions.[34]

The exchange of counterion B on a resin for counterion A in a well-agitated solution can be considered as taking place in three distinct steps:

Step 1. Transfer of A from deep in the solution phase to the vicinity of the resin. In a well agitated system this transfer is by convective transport, it is very rapid, and it is not rate controlling.

Step 2. Transport of counterions across a thin film (the Nernst film) surrounding the ion exchange particle

Step 3. Interdiffusion of ions A and B within the ion exchange particle itself.

The Nernst film is an approximation for treating the complex problem of mass transfer in the external phase.[6] Boyd and co-workers were the first to apply the Nernst film concept to ion exchange.[35] In an agitated system of a solid and a liquid, convection is present throughout the liquid phase and absent in the solid. Moreover, convection, and hence mass transfer by convection, diminishes gradually as we approach the solid–liquid boundary. To simplify treatment of such a system, Nernst introduced the idealization of considering the stirred liquid phase as comprising two zones: *Zone 1* is a bulk liquid phase up to a distance δ from the liquid–solid interface. In this zone mass transfer is by convection and is rapid. *Zone 2* is a thin film adherent to the solid through which mass transfer takes place by diffusion and is relatively slow. This zone is the Nernst film.

Using the Nernst approximation, the effect of agitation is seen as causing a thinning of this hypothetical layer. The thickness cannot be reduced indefinitely, however, but appears to reach a limit of about 5×10^{-4} cm.[6] Quite large values of δ are to be expected for poorly agitated or viscous systems.

Any real ion exchange process will involve a combination of the two diffusional steps but it has been instructive for both theoreticians and experimentalists to study systems that represent limiting cases, that is, solely particle diffusion control and solely film control. To further simplify a very complex problem it has been helpful to consider the exchange of *isotopic* species between the resin and solution, and the use of radiolabeled species has further facilitated the experimental checking of many of the theoretical models. We will examine two rate expressions that describe the rate of isotopic exchange between a quantity of ion exchange spheres and an "infinite" bath of solution.[6]

For a reaction that is controlled purely by diffusion within the particle, the fractional attainment of equilibrium, F, as a function of time, t, is given by

$$F = 1 - (6/\pi^2) \sum_{n=1}^{\infty} (1/n^2) \exp\left(-D_R t \pi^2 n^2 / r^2\right) \qquad (4.49)$$

where r is the radius of the particle and D_R is the self-diffusion coefficient for the exchanging ion in the resin phase. The half time of exchange may be calculated from (4.49); the substitution $F = 0.5$ gives

$$t_{1/2} = 0.03r^2/D_R \qquad (4.50)$$

Note that in this case the half time depends only on the size of the particle and the diffusion coefficient within the particle; it is independent of the concentration of the external solution.

For a reaction that is controlled solely by diffusion in the boundary film

$$F = 1 - \exp(-3D_S C_S t/r\delta C_R) \qquad (4.51)$$

where D_s is the self-diffusion coefficient for the ion in the solution phase, C_s is the concentration of the ion in the solution phase, δ is the thickness of the Nernst film, and C_R is the capacity of the resin. For this case the half-time is given by

$$t_{1/2} = 0.023r\delta C_R/D_S C_S \qquad (4.52)$$

The two rate-controlling regimes may be compared by examining plots of half-times as a function of particle radius (Figures 4.11 and 4.12). For a relatively large particle ($r = 0.1$ cm) and a "fast" ion ($D = 10^{-6}$ cm^2 s^{-1}) the half-time for the particle control is about 300 s. For film control from a 0.1 M solution

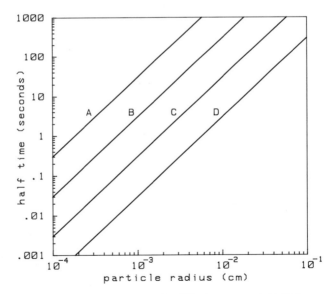

FIG. 4.11. Ion exchange kinetics—particle diffusion control. Half-times of exchange were calculated from equation (4.50). The plots represent different diffusion coefficients (D_R) in the resin phase (A, 10^{-9}; B, 10^{-8}; C, 10^{-7}; and D, 10^{-6} cm^2 s^{-1}).

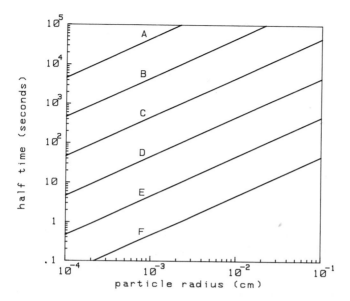

FIG. 4.12. Ion exchange kinetics—film diffusion control. Half-times of exchange were calculated from equation (4.52). The plots represent different concentrations in the solution phase (A, 10^{-6}; B, 10^{-5}; C, 10^{-4}; D, 10^{-3}; E, 10^{-2}; and F, 10^{-1} M). The Nernst film thickness was assumed to be 10^{-3} cm. The diffusion coefficient in the solution phase was assumed to be 10^{-5} cm^2 s^{-1}.

the half-time is between 40 and 50 s, so we would conclude that under these circumstances the slow reaction is within the particle. For a 10-μm-diameter particle, on the other hand, the half-time for film control is between 0.2 and 0.3 s, while for particle control it is about 0.08 s. Further comparisons of this type would reveal that conditions that favor particle diffusion control are large particle size, high concentration in the external phase, and low diffusivity in the resin phase. Conditions that favor film control are small particle size and low concentration in the external phase.

While diffusion coefficients in the external phase generally fall in the range of 10^{-5}–5×10^{-5} cm^2 s^{-1}, diffusion of ions in resins can be extremely slow by comparison, especially for highly cross-linked resins (Figure 4.13) and if the ions are bulky.

The rate laws for mass transfer in ion exchange as exemplified by (4.49) and (4.50) are obviously complex in form and derivation. Incorporating them into a description of chromatographic column behavior presents problems of daunting mathematical complexity. As a result, most chromatographers, while they may have a good theoretical understanding of the equilibrium aspects of separations, rely on an experimental/empirical approach in assessing kinetic effects. A desir-

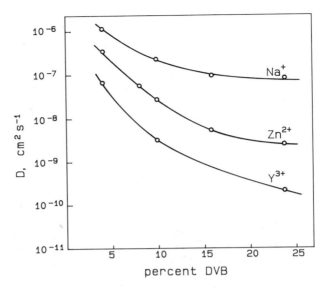

FIG. 4.13. Self-diffusion coefficients at 25°C for Na^+, Zn^{2+}, and Y^{3+} in sulfonated styrene–DVB resins. (From Ref. 36 with permission.)

able compromise might be a way of roughly estimating the relative importance of the various kinetic factors; this is discussed in the following section.

4.3.1. Band Broadening in Ion Exchange Chromatography

Separation in ion exchange chromatography as in other forms of chromatography depends on two factors: disengagement of the eluite bands and the dispersion of the bands. Previous sections of this chapter have dealt mainly with the first of these; we will now consider band broadening.

Band broadening in ion exchange chromatography, as in other forms of chromatography (Section 2.4.4), depends on three factors: (1) longitudinal diffusion, (2) eddy diffusion, and (3) mass transfer within the stationary and mobile phases. Glueckauf[37,38] was one of the earliest to attempt a detailed description of the ion exchange process. It took the form of the equation that bears his name and relates theoretical plate height, H, to other parameters of the system:

$$H = 1.64r \qquad \text{Term I}$$
$$+ [K_D/(K_D + \alpha)^2] [0.142r^2 \, F/D_S] \qquad \text{Term II}$$
$$+ [K_D/(K_D + \alpha)^2] \{0.266r^2 \, F/[D_L (1 + 70 \, r \, F)]\} \qquad \text{Term III}$$

where K_D is the distribution factor, α is the void fraction of the column, F is the flowrate, r is the radius of the particle, D_S is the diffusion coefficient of eluite within the exchanger particle, and D_L is the diffusion coefficient of eluite in the liquid (mobile) phase.

The first term is the contribution from eddy diffusion, the second the contribution from mass transfer resistance within the particle, and the third the contribution from mass transfer in the mobile phase. Band dispersion due to longitudinal diffusion is ignored as being insignificant under practical chromatographic conditions. The equation reflects the usual expectations of chromatography, namely, that peak sharpness (i.e., low HETP) is favored by operating at low flow rates and with small particle size.

The Glueckauf equation was developed for spherical packings with functionality throughout the bead and is not directly applicable to much of ion chromatography, which uses materials that have active sites only on a thin outer shell. Shell functionalized materials offer an advantage for chromatography in that they limit diffusion to this thin region on the surface of the stationary phase that should lead to more efficient column behavior. On the other hand, the thinner the shell the less is the capacity per unit volume of column. Therefore the development of "pellicular" or "superficial" exchangers requires optimization between these two conflicting factors. Hansen and Gilbert[39] derived a modification of the Glueckauf equation applicable to shell-type exchangers and providing a means for optimizing the thickness of the shell. Stevens and Small,[40] in a practical study of surface sulfonated resins, found a considerable discrepancy between their experimentally determined optimum and the theoretical value as predicted by Hansen and Gilbert. The measured optimum thickness was much lower than the theoretical value of Hansen and Gilbert and they concluded that abnormally low diffusivities in the sulfonated layer might be a reason for the disagreement.

Compared to work on ion exchange equilibria and selectivity, there has been relatively little published work on optimizing column efficiency in IC. Much of this type of research is carried out by the column manufacturers, and so details tend to remain proprietary.

Most theoretical work on band broadening tends to be overly complex for the average chromatographer and is usually ignored. A more intuitive treatment, if lacking in mathematical rigor, could be more useful. Furthermore, in many instances it is probably enough that the user develop a "feel" for the relative magnitude of the factors that control column efficiency. Combining in a simple way what is known of resin kinetics and column rate theory offers one such approach.

Recall from Chapter 2 that the general expression for the height equivalent to a theoretical plate, H, is given by

$$H = 2D/U + d_p + 2P(1 - P)U\tau_S \tag{4.53}$$

The first term, the contribution from longitudinal diffusion, can usually be ignored in current IC practice, so it is left then to assess the relative importance of the other two terms.

Equation (4.53) gives the general form of the relationship, but in a more rigorous mathematical treatment each term must be modified with an appropriate coefficient.[41,42] The coefficient for the second term has not been satisfactorily established, but it is a reasonable approximation to assume that the theoretical plate height in a column can never be much less than the diameter of the particles and so we will let the coefficient be unity. For spherical ion exchange particles the coefficient for term 3 is 0.133.[42]

There remains the problem of assessing the relaxation time τ_S. If diffusion in the ion exchange particle is rate controlling then a reasonable value for τ_S is the average time an eluite ion would take to diffuse a distance equal to the radius of the particle. This can be calculated using the Einstein equation for diffusion

$$\tau_S = r^2/2D_R \qquad (4.54)$$

where D_R is the diffusion coefficient in the resin. Substituting in equation (4.53), omitting term 1, and introducing the modifying factor for term 3 gives

$$H = d_p + 0.133P(1 - P)Ur^2/D_R \qquad (4.55)$$

Now let us examine some test cases. In the first case we will assume a 20-μm-diameter particle, $P = 0.5$, a linear velocity of 1 cm/s, and a relatively fast diffusion in the resin ($D_R = 10^{-6}$ cm^{2}s^{-1}). In this instance term 2 has the value of 0.033 cm, which would suggest that mass transfer is an important if not the dominant factor since term 1 in 4.55 is only 0.002 cm. If, on the other hand, we assume a pellicular ion exchanger with a shell thickness of only 0.2 μm then the second term calculates to be only 3×10^{-5} cm. In this case it would be reasonable to attribute column efficiency mainly to packing effects in the column rather than mass transfer resistance in the resin.

Mass transfer resistance in the mobile phase is another potential source of band broadening. With fully functionalized particles it is unlikely to be important since (1) the average diffusional distance in the resin is about six times greater than in the mobile phase for a packed bed of spheres, and (2) diffusion coefficients are at least ten times lower in the resin than in solution. For pellicular-type resins, however, the diffusional distance in the stationary phase can now be considerably less than in the mobile phase. Then it seems appropriate to apply term 2 to the mobile phase, where r is made equal to the average diffusional distance in the void element. For an assembly of spheres this is approximately 15% of the diameter of the spheres in the assembly. For D_R we substitute instead a value for diffusion in a liquid; a value of 10^{-5} cm^2 s^{-1} is suitable. Assuming then a 20-μm-diameter substrate particle the mean diffusional distance in the mobile phase is approximately 1.5×10^{-4} cm. Under these conditions term 2 has a value of 7.5×10^{-5} cm. Since this is still considerably less than term 1 we conclude that mass transfer in the mobile phase does not determine plate height. From these calculations we draw the general conclusion that for the exchange of small, fast ions on pellicular resins the most important factor in determining column performance is proper packing of the column. For more sluggish ions where diffusivities in the resin phase could be as much as 100 times less, the mass transfer in the resin can be a significant factor. Under these conditions we would then expect efficiency to have a marked dependence on flow rate as predicted by the second term of (4.53).

4.3.2. Chemical Rate Control

The rate of ion exchange reactions is in most cases diffusion controlled. Riviello, however, has observed a variation in column efficiency that he attributes to a slow reaction in the mobile phase.[34] The system he studied involved the separation of transition metal ions using a surface sulfonated resin and oxalate ion as complexing agent. He observed a marked difference in HETP for the elution of the various ions (Table 4.8) when oxalate was used. In a noncomplexing medium such as sodium perchlorate, there are differences but they are small by comparison. Riviello found a correlation between HETP and the rate constants for exchange of water between the inner coordination sphere of the ion and bulk water—the higher the rate constant the lower the plate height for that ion (Figure 4.14). Noting the parallel correlation between the rate of water exchange and the rate of complexation he argued that the relatively poor efficiency as shown by nickel, for example, was due to the sluggishness of the complexation reaction. If this is slow relative to the diffusion then it becomes the rate-determining step and leads to less efficient columns than when complexing is absent.

TABLE 4.8. Comparison of HETP Values for Metals[a]

| Metal ion | k' | HETP (mm) | |
		$NaClO_4$	Na_2Ox
Mn^{2+}	4.2	0.0873	0.0962
Fe^{2+}	4.2	0.0951	0.0975
Co^{2+}	4.1	0.0973	0.103
Ni^{2+}	4.2	0.0809	0.692
Cu^{2+}	4.3	0.104	0.0891
Zn^{2+}	3.9	0.0864	0.0926
Cd^{2+}	5.1	0.0843	0.101

[a]Reference 34.

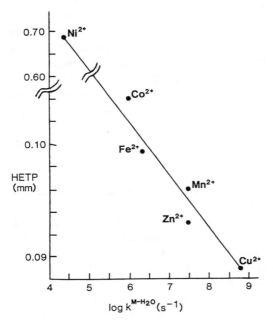

FIG. 4.14. HETP versus rate constant for water exchange (k^{M-H_2O}) for the separation of transition metals on a surface sulfonated resin. Eluent: 0.015 M oxalic acid adjusted to pH 4.2 with NaOH. (From Ref. 34 with permission.)

4.4. ION EXCHANGE CHROMATOGRAPHY

There are three main ways in which ion exchange chromatography may be performed:

1. Elution development.
2. Frontal analysis.
3. Displacement development.

Of the three methods, it is the first that is employed in ion chromatography.

4.4.1. Elution Development

In elution development the stationary phase is first equilibrated with mobile phase. A small amount of sample is then introduced into the flowing mobile phase, at which time chromatographic sorting of the sample components begins. Eventually the individual components are eluted in the order of their distribution coefficients, the chromatogram having the form of a series of peaks with a roughly Gaussian shape. In elution development a successful separation is characterized by completely separated peaks, which is one of the reasons for its appeal as an analytical technique.

In most other forms of chromatography one is exclusively concerned with the concentration profiles of the eluites in the effluent; it is rarely that one is interested in the fate of the eluent species. In IC, on the other hand, the changes in the concentration of the *eluent* ion(s) are sometimes of at least equal interest, for it is the concerted changes in the concentrations of eluite and eluent that control the detectability of the analytes in many cases. Since this feature of elution development is of most relevance for detection mechanisms, further discussion will be taken up in sections devoted to detection.

4.4.2. Frontal Analysis

In frontal analysis a mixture of components is fed continuously to a column, in which case the effluent history is characteristically a series of breakthrough curves. Figure 4.15A illustrates a hypothetical ion exchange case. An ion exchanger in the counterion form A receives a continuous feed containing ions B, C, and D that are of the same charge as A. If the selectivity order is D > C > B > A then the concentration of each in the effluent will be as shown in Figure 4.15B. Note that the total concentration of all the species in the effluent at any time is constant and equal to the influent concentration. (In actual practice this is not exactly so since gain or loss of water by resins as they change ionic forms can concentrate or dilute the mobile phase. Ordinarily this is a minor effect.)

Frontal analysis is sometimes used to produce a pure component from a mixture, but it is of limited utility. In the hypothetical separation of Figure 4.15 the only component of the mixture B, C, D that can be obtained in pure form is B. Frontal analysis as a technique is of little practical utility in IC, but as a phenomenon it does have some practical implications.

4.4.3. Displacement Development

Displacement development as practiced in ion exchange chromatography is a powerful means of separation, particularly on a preparative scale. In displacement development a limited but relatively large fraction of the ion exchange bed is loaded with sample B, C, D (Figure 4.16). This band of ions is then displaced forward in the bed by ion E in the eluent and eventually becomes segregated into contiguous bands of the separate components with the ordering again dictated by the selectivity of the system—ions with lower affinity for the resin moving to the downstream side of the band, higher-affinity ions to the upstream side. For this method to work best, the order of selectivities should be E > (B, C, D) > A; this ensures sharpness at the leading and trailing boundaries of the band. Within the band itself the boundaries between contiguous components are naturally self-sharpening (see Section 4.4.4). There are means of calculating the minimum amount of bed necessary to bring about a complete segregation of the components.[43]

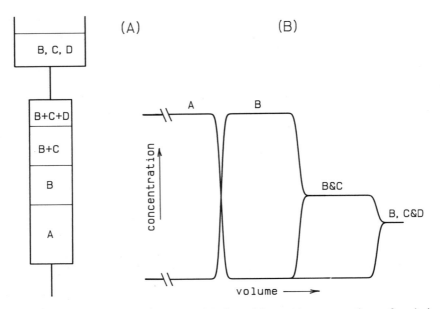

FIG. 4.15. Frontal analysis. A mixture of B, C, and D, equal concentrations of each, is loaded continuously onto a resin in the A form. Plot shows the effluent concentrations of the several ions as a function of effluent volume. The volumes of the various bands will depend on the relative affinities of the ions for the ion exchanger.

The large-scale production of pure rare earth elements owes much to ion exchange displacement development and the pioneering work of Spedding and collaborators.[44] In a typical separation scheme, a mixture of lanthanide ions is loaded onto the cupric ion form of a strong acid cation exchanger (Figure 4.17). The greater affinity of the resin for the trivalent lanthanide over the divalent copper assures a sharp downstream boundary at this stage (see Section 4.4.4). After loading, the band of lanthanide ions is displaced by a solution containing Na_3HEDTA. At the upstream sodium–lanthanide boundary lanthanides are removed:

$$(R-SO_3^-)_3M + Na_3HEDTA \rightarrow 3R-SO_3^-Na^+ + MHEDTA$$

Because of the high stability of the lanthanide–EDTA complexes this reaction is very favored and the boundary remains sharp. The lanthanide complexes then percolate through the band and it is in this step that sorting takes place:

$$(R-SO_3^-)M_I + M_{II}HEDTA \rightleftharpoons (R-SO_3^-)_3M_{II} + M_IHEDTA$$

with the lanthanides with the greater complex stabilities moving to the downstream side of the band and the less stable to the upstream. When the lanthanide

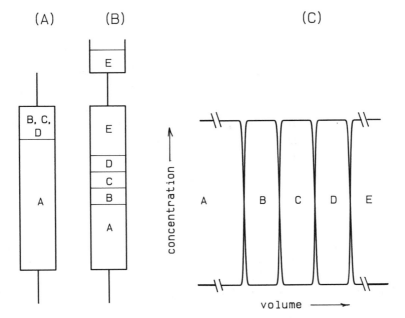

FIG. 4.16. Displacement development. A mixture of B, C, and D is loaded onto a resin in the A form. The sample band is then displaced by E and segregates into contiguous zones in an order determined by the relative selectivities of the ions with the least tightly held ion moving to the front of the band and the others following in the order of ascending affinity. The widths of the B, C, and D bands depend on their relative amounts in the sample loaded.

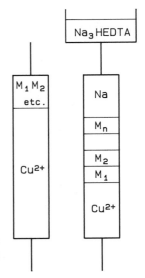

FIG. 4.17. Displacement development separation scheme for lanthanide ions using Cu^{2+} form resin as the retaining resin and sodium EDTA as the displacing electrolyte.

EDTA stream reaches the copper form resin the lanthanide species are effectively "forced onto" the resin because of the much greater stability of the copper–EDTA complex and the downstream boundary remains sharp as a result.

Displacement development is largely used as a preparative tool in ion exchange; it has not been explored to any significant extent in an analytical context. Even in a very successful separation by displacement development there is not complete disengagement of the components; there is always a slight overlap between contiguous zones. It is this feature that undoubtedly lowers its appeal as an analytical procedure.

4.4.4. Peak Shape in Elution Development

Resolution of peaks in IC is determined by peak width and peak shape as well as by selectivity. As we have seen earlier, peak width is mainly determined by kinetic and packing effects in the column, while peak asymmetry is more likely due to equilibrium effects.

Eluite peaks in IC are as a rule approximately symmetrical and Gaussian but can with some species and under some circumstances be quite asymmetrical. Overloading a column invariably leads to asymmetry. If the amount of eluite is sufficiently small then elution takes place in a condition of constant K_D and symmetrical peaks are obtained. Under this condition peak height is proportional to the amount of sample loaded. As the amount of sample is increased, a point is reached where the peaks become asymmetrical and the proportionality between peak height and sample amount breaks down. The causes of this may be illustrated by considering the displacement of a band of eluite S by eluent ion denoted E (Figure 4.18). To facilitate the argument it will be assumed that in elution development all eluites start off as a plug-shaped band, albeit a very narrow one. Figure 4.18 therefore represents the eluite displaced but a short distance into the bed. We will first consider the case where the resin prefers S, and later where it prefers the eluent ion.

Consider the movement of zone 1 relative to zone 2 at the E/S upstream boundary. Zone 1 is representative of regions where a trace of S is moving in a preponderance of E while for zone 2 it is the reverse—a trace of E being eluted by S. If the resin prefers S then the k-prime of zone 1 will be greater than that of zone 2 and they will tend to move apart. This will cause the upstream boundary to become more diffuse as elution proceeds. By exactly similar reasoning zone 3 will tend to catch up with zone 4 so the downstream boundary is self-sharpening. The diminishing sharpness of the upstream boundary as elution proceeds causes S to disperse and to reach trace levels throughout the S band, at which point all regions of the band will move with a constant k-prime characteristic of the elution of S by E. When the amount of eluite loaded is small, this condition will be reached after but a short time into the elution and one would not expect the

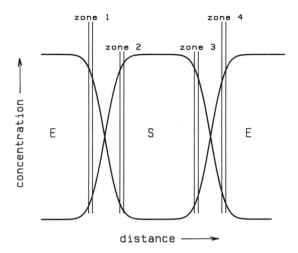

FIG. 4.18. Concentration profile of sample ion S and eluent ion E, within an ion exchange column, as a function of distance from the column inlet. Illustrates a sample overload situation.

peak shape to be affected by the amount of S loaded. On the other hand, if the exchanger is overloaded, some residue of the original band shape may remain when the eluite eventually elutes from the column, and we would expect the proportionality between peak height and eluite amount to be violated.

The direction of bias in an overloaded peak, that is, whether it is diffuse at the upstream or downstream side, will depend on the relative affinities of S and E. If S is the more tightly held of the two then the peak as seen by the detector should have a sharp front and a diffuse tail—a condition normally referred to as tailing. If E is the preferred ion then the opposite effect, sometimes called fronting, would be expected.

There are instances of peak asymmetry in IC that are not readily explained by overloading. For example, in anion IC using carbonate eluents the peak due to nitrate ion invariably tails even at levels where other ions of similar k-prime give symmetrical peaks. A plausible explanation for this presumes the resin to contain a few sites with a much higher affinity for nitrate than the majority—sites in more hydrophobic regions of the resin, for example. In this model, since there are but few of these high-affinity sites, they will tend to overload with even quite small amounts of sample and show symptoms of overloading. When carbonate eluents are replaced with a more hydrophobic species such as cyanophenol, nitrate gives symmetrical peaks. The more hydrophobic cyanophenate is presumed to compete for the high-affinity sites more effectively than carbonate.

4.4.5. Base-Line Disturbances, "System" Peaks

In certain ion exchange separations peaks or negative disturbances appear in addition to those due to the eluite ions. Since they are not attributable to eluites from the sample they are often referred to as system peaks. They have been observed when using conductometric detection and indirect photometric detection and particularly when using weak organic acid salts as eluents. They can be a major source of interference if their elution coincides with an eluite of interest, so it is important to understand how they arise.

We will begin by considering a simple ion exchange chromatographic system using an eluent with but a single displacing ion denoted A. It will be further assumed that the exchanger in counterion form A has been equilibrated with eluent containing A at concentration denoted c. When a sample containing a single eluite species S is injected there is an exchange of S for A on the column, and a pulse of displaced A with its co-ion passes rapidly through the column and exits at the void volume of the column. If the detector is monitoring a property that is possessed to any extent by A, then a base-line disturbance will usually be observed. This is called the void disturbance. It can be in a positive or negative direction depending on whether the concentration of released A is greater or less than the ambient concentration in the eluent. If the concentration of the sample is greater than the eluent then it will be positive and vice versa. If the concentrations of S and A are the same then, theoretically, no void disturbance should occur. Some time after the void disturbance the sample eluite will elute and cause a perturbation in the base-line signal. For this simple system these are the only sources of base-line disturbance; after S has passed out of the column the system has reverted to its original condition.

Consider now a system of an exchanger equilibrated with an eluent containing two displacing ions A and B. When the exchanger is equilibrated with the A/B mixture, and assuming that A and B are both monovalent, then the composition of the exchanger and solution phases are related by the familiar mass action expression

$$[B]_R/[B]_S = K_A^B [A]_R/[A]_S \tag{4.56}$$

The quantities $[B]_R/[B]_S$ and $[A]_R/[A]_S$ are the distribution coefficients of the B and A species, respectively, and they are related through the selectivity coefficient of the exchange reaction.

Now let us examine the effects to be expected from various types of injection into this system.

Case 1: An injection of a mixture of A and B where the ratio of A to B is the same as that in the eluent but the total concentration differs from that in the eluent. In this case we would expect but a single disturbance at the void since according to (4.56) the composition of the exchanger should remain unaltered by the passage of the A/B "slug"; the composition of the exchanger is dependent only on the ratio of A to B and not on solution concentrations. (Strictly speaking this is true only in an ideal system where selectivity is independent of solution concentration. For mono-mono-valent exchange this is approximately true.)

Case 2: An injection of sample eluite S. Here we can expect at least one disturbance, at the elution position of S, and a second at the void if the sample concentration differs from the concentration of the eluent.

Case 3: An injection of a mixture of A and B where the ratio of A to B differs from that in the eluent. Let us assume that the injection is richer in B than is the eluent and furthermore that the exchanger prefers B over A ($K_A^B > 1$). This will cause a change in the exchanger composition at the column entrance so that it becomes richer in B. The injected slug will in turn become equilibrated with the exchanger within a short distance into the bed and will acquire the ratio of A to B of the eluent and pass out in the void, perturbing the base line if the slug concentration differs from that of the eluent. But still left in the column is a B-rich zone, which will now elute at a rate determined by the distribution coefficient of species B. This will eventually elute from the column and will cause a positive disturbance if the detector is sensitive to B. If the detector is sensitive to A and not to B there will be a negative disturbance since a rise in B must be accompanied by a fall in the concentration of A. This shows how a disturbance of the base line can arise not by injection of a sample but by a disturbance of the eluent composition.

In normal chromatographic procedure there is no reason to deliberately disturb the eluent in this way, but are there ways in which it can be done inadvertently? Consider a case where B is a divalent ion and A is monovalent. Since the equilibrium depends not only on the ratio of A to B but also on *total solution concentration* [see (4.18)] then sample injection can lead to two disturbances besides any void disturbance that may occur. The source of one disturbance is obvious—that due to eluite S—but the other is of more subtle origin. Let us suppose that the concentration of the injected sample is less than the eluent concentration. Sample S will displace A and B into the injected slug but A and B are now in an environment (lower concentration) that is more favorable for B on the exchanger so the slug will tend to "dump" some B in preference to A onto the exchanger thus creating a zone that is B-rich relative to that upstream or downstream. The slug itself will acquire a new A/B ratio that is compatible with its total concentration and with the composition of the exchanger downstream, and pass out in the void. The eluent will then proceed to displace the B-rich zone eventually causing a perturbation when it elutes from the column. Are there situations in IC that compare to this hypothetical case? It is commonplace in IC for sample concentration to differ from the eluent concentration; indeed, it could fairly be said to be the rule. Eluents comprising monovalent and divalent mixtures are widely used; carbonate–bicarbonate in suppressed conductometric IC and phthalate–biphthalate in nonsuppressed conductometric and indirect photometric detection are well known examples. In the case of suppressed conductometric methods base-line disturbances other than those due to analyte tend to be eliminated by the suppressor, but in the other cases disturbances are to be expected and have been observed.

Concentration of the sample is not the only factor that can elicit a system disturbance—the pH of the sample can also be critical. For example, an anion exchanger equilibrated with a phthalate–biphthalate eluent can be expected to enrich itself in phthalate if a high pH sample is injected. This excess phthalate will slowly elute and is likely to cause a disturbance in a photometric detector as it leaves the column.

4.5. ION INTERACTION CHROMATOGRAPHY

Ion interaction chromatography (IIC) is one of several names used to describe those IC methods that use a neutral, nonpolar stationary phase in conjunction with an ion interaction reagent (IIR) in the mobile phase (see Chapter 3). There are two types of stationary phase in common use: the silica-based, bonded phase packings of reverse phase chromatography and organic polymer particles based on polystyrene. In either case the stationary phases are permeated by a network of fine pores that give them an extensive interface with the mobile phase, typically on the order of several hundred square meters per gram of stationary phase.

The IIR is usually, but not always, a lipophilic ion. It is always opposite in charge to the analyte species being examined, and for this reason is sometimes referred to as the counterion.[45,46] I will not use that term to describe an IIR since doing so can lead to confusion when drawing parallels between ion interaction chromatography and ion exchange chromatography, where the term "counterion" has a quite different meaning.

When the analytes are anionic, the IIRs used are typically surface active quaternary ammonium compounds. When they are cationic then surface active anions such as alkyl sulfates or sulfonates are employed. Should the analytes themselves be surface active then the IIR will usually be an ion that would not normally be considered to be surface active—chloride or perchlorate, for example, in the chromatography of cationic surfactants, ammonium ion in the case of anionic surfactants.

Normally some IIR must be maintained in the mobile phase in order for the system to perform reproducibly but there are cases of satisfactory performance where the nonpolar phase is pretreated with IIR but the IIR is omitted from the eluent.[47]

In addition to the IIR and water, the mobile phase will usually contain a quantity of another electrolyte and often some water-miscible solvent such as methanol or acetonitrile. This additional organic solvent is sometimes called the modifier.

In a typical chromatographic operation of this type, mobile phase consisting of water, modifier, IIR, and electrolyte is passed through a bed of stationary phase for such time as is necessary to bring the two phases into equilibrium. The system is then ready to perform analyses.

Several mechanisms have been proposed to account for the retention behavior of eluites in ion interaction systems. An early theory proposed that the lipophilic ion formed neutral ion pairs with analyte ions, which then distributed between the mobile and stationary phases. Differences in retention of analytes were attributed to differences in their tendency to form ion pairs.[48] This was later disputed on the grounds that conductivity measurements indicated little if any association between ions in the mobile phase.[49] That is not to say, however, that ion pair formation does not occur at the stationary–mobile phase interface, where the lower dielectric properties of this environment should favor electrostatic association of ions of opposite charge.

At the present time the most accepted model of ion interaction systems stresses the analogy with ion exchange while recognizing the influence of surface chemical effects in determining some of the fundamental properties of these "pseudo-ion-exchangers." We will examine this model as it might be applied to the chromatography of small organic or inorganic anions in the presence of a lipophilic cation, Q^+, in the mobile phase.

4.5.1. The Electrical Double Layer

We have noted that Q^+ is typically a quaternary ammonium ion $R_1R_2R_3R_4N^+$. The R-groups can comprise a mixture of chain lengths, some of them frequently as high as C_{18} or greater, but sufficiently bulky in total so that Q^+ is lipophilic (hydrophobic) and tends to adsorb at the interface between the nonpolar stationary phase and the essentially aqueous mobile phase. Carried with Q^+ to this interface to maintain approximate electroneutrality, is an equivalent number of ions of opposite charge. The bulk of this counterion layer is closely associated with the adsorbed Q^+ layer but a minutely small amount occupies a diffuse region extending into the mobile phase. The interfacial sorption of surface active ions, and the disposition of charge in the counterion layers about a charged surface, are classical problems of surface chemistry.[50] Cantwell and co-workers were among the first to develop the relationship between IIC and the theories of electrical double layers, and a reading of their and other similar treatments[46,51] is recommended. The following extension of the ion exchange analogy introduces these surface chemical features in a qualitative way.

Assuming that the conditions in the mobile phase are fixed, equilibration of the two phases will eventually lead to adsorption of a certain level of Q^+ at the interface along with a closely associated layer of anions, A^-, of equivalent charge. Q^+ is analogous to the functional group on an ion exchanger except that in the case of IIC it is sorptively bound to the matrix, while in an ion exchanger the attachment is covalent. The amount of Q^+ sorbed under a fixed set of conditions is analogous to the capacity of an ion exchanger. The counterions, A^-, are free to exchange with others of similar sign in the mobile phase, which includes eluite ions.

Of the many properties and parameters of an ion exchange system, one of the most important in determining eluite retention is the ion exchange capacity. Analogously, one would therefore expect the "capacity" of the double layer to exert a comparably important effect in IIC systems. There is a marked difference, however, in the two systems with respect to what might be called the "stability" of their respective capacities.

For ion exchangers of the strong acid or base types, capacity is totally stable against a variety of changes in the mobile phase. For example, changing the counterion or its concentration or adding another solvent to the mobile phase has no effect whatsoever on the capacity of an ion exchanger. In contrast the adsorption of Q^+ and, as a consequence, the capacity of the double layer can be profoundly affected by changes in the composition of the mobile phase as well as by the nature of Q^+ itself. We will now examine the sorption process in some detail, focusing on the capacity of the double layer, on how it changes, and on what implications this has for chromatography.

4.5.2. The "Capacity" of the Adsorbed Double Layer

The adsorption of Q^+ onto the stationary/mobile phase interface depends on the following factors: the concentration and nature of Q^+, the concentration and nature of A^-, the concentration and nature of any modifier added to the system, and the temperature of the system.

4.5.2a. Q^+ and Its Concentration

Naturally as the concentration of Q^+ increases in the mobile phase, so too does its concentration at the interface, and this continues until the interface is saturated with lipophile. Adsorption isotherms express the dependence of double layer capacity on the concentration of IIR, but such data are generally lacking for IIC systems. The extent of adsorption of a charged lipophile is, as a rule, less than that of an uncharged species of similar structure. This is due to electrostatic repulsion effects from the double layer. Because of the diffuse nature of the counterion layer, the positive charge of the adsorbed Q^+ layer is not completely shielded and the sorption of Q^+ is opposed by electrostatic repulsion from the growing positive layer. The position of equilibrium is determined, among other factors, by the relative strengths of the hydrophobic interaction promoting adsorption and the electrostatic forces opposing it. We would therefore expect the amount of Q^+ adsorbed to increase with increasing lipophilicity; the more hydrophobic Q^+ the higher the capacity of the double layer, all other factors being equal. If the concentration of Q^+ is high enough or if Q^+ is sufficiently hydrophobic the interface will reach a point of saturation with adsorbed lipophile, at which point the differences between various lipophiles with respect to double layer capacities diminishes.

4.5.2b. A^- and Its Concentration

The adsorption of IIRs, and consequently the capacity of the double layer, are strongly dependent on the counterion A^- and its concentration. Cantwell and Puon[51] studied the adsorption of the diphenylguanidinium ion onto macroporous polystyrene in the presence of sodium chloride; the results are illustrated in Figure 4.19. It is evident that increasing the concentration of salt increases the surface "capacity" of the adsorbed layer at a fixed level of DPGH$^+$ in the aqueous phase. This is a result of the suppression of the repulsive interaction by the added electrolyte or, expressed in another way, a more effective shielding of the adsorbed layer, which permits more adsorption of the surface active cation. Consider now what implications this effect has for eluite retention in an IIC system as compared to an ion exchange system.

Increasing the electrolyte level in the mobile phase will in both cases oppose eluite retention simply because of ion exchange competition between eluite and eluent ions for the counterion layer. But in the IIC case the "capacity" increases with added electrolyte (Figure 4.19), while in the ion exchange case the capacity remains constant. Thus in the IIC case the rising capacity counteracts somewhat the increasing concentration of the displacing ion in determining the rate of eluite displacement. The dependence of k-prime on mobile phase electrolyte concentration is therefore a more complicated relationship for IIC systems than it is for ion exchange systems where the relationship is comparatively simple.

Cantwell and Puon[51] studied the adsorption of lipophilic ions on porous polystyrene and explained their observations in terms of the Stern–Gouy–Chapman (SGC) theory of the electrical layer. In a chromatographic system comprising mixtures of tetrabutylammonium bromide plus added bromide as mobile phase and porous polystyrene as stationary phase they studied the elution of injections of low concentrations of DPGH$^+$. They observed a dependence of k-prime on the concentration of bromide added that agreed with their theoretically derived relationship

$$1/V = 1/k_5 + 1/(k_4 C^{1/2})$$

where V is the elution volume of the DPGH$^+$, C is the concentration of bromide, and k_4 and k_5 are constants.

Although this experiment was not designed to measure the adsorption of the majority lipophile (TBA$^+$), which was present in excess, we can nevertheless

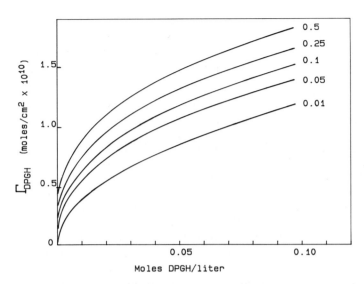

FIG. 4.19. Adsorption of diphenylguanidinium ion onto a macroporous polymer. Adsorption isotherms for DPGH$^+$ in the presence of various bulk solution concentrations of sodium chloride. The numbers at the end of the curves are the molarities of sodium chloride. (From Ref. 51 with permission. Copyright 1979 American Chemical Society.)

reasonably assume that the behavior of the minority lipophile is also a reflection of the behavior of the TBA$^+$. (An experiment in which radiolabeled TBA$^+$ replaced DPGH$^+$ would serve the purpose better.)

Iskandarani and Pietrzyk[46] have studied the retention of inorganic analytes (NO$_2^-$ and NO$_3^-$) in an IIC system consisting of porous poly(styrene–DVB) as stationary phase. The mobile phase IIR was tetrapentylammonium ion at a fixed level and sodium fluoride at varying levels. Figure 4.20 shows a plot of $1/k'$ versus the concentration of fluoride displacing ion, and according to these authors the dependence of eluite retention on eluent concentration is consistent with double layer theory. The degree of dependence differs considerably from that of an ion exchange system. There, a tenfold change in eluent concentration would lead to a tenfold change in k-prime for mono-mono-valent ion exchange, whereas in the IIC system the change is only threefold for nitrite ion. And for nitrate it appears to be even less than this.

The effect of the counterion A$^-$ on the sorption of Q$^+$ has been studied by a number of workers.[46,49] Figure 4.21 shows the effect of counterion type on the adsorbability of benzylammonium ion onto PSDVB. Cantwell and Puon advance a number of possible explanations for the differences between the various counterions; greater association or ion pair formation between Q$^+$ and counterion leading to better shielding of the positive layer is a plausible one. It is

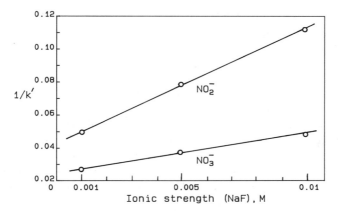

FIG. 4.20. Retention of inorganic ions on a macroporous polymer (Hamilton PRP-1) as a function of eluent (NaF) concentration. Mobile phase conditions: 10^{-3} M tetrapentylammonium fluoride and NaF, in acetonitrile–water (15 : 85 volume ratio). (From Ref. 46 with permission. Copyright 1982 American Chemical Society.)

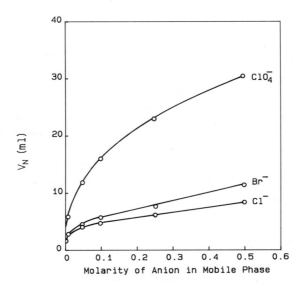

FIG. 4.21. The effect of counterion type in the mobile phase on the adjusted retention volume of benzylammonium ion on a macroporous polymer (Rohm and Haas XAD-2). (From Ref. 51 with permission. Copyright 1979 American Chemical Society.)

especially noteworthy that, in this as in other work, the order of efficacy is the same as the order of affinities in an ion exchange system; that is,

$$ClO_4^- > I^- > NO_3^- > Br^- > NO_2^- > Cl^- > F^- > OH^-$$

While the bulk ion has a major effect on the sorption of Q^+ the eluite ion also exerts a local effect as it elutes through the bed of stationary phase. Hackzell and Schill[52] have shown that if the eluite ion has a greater specific binding to Q^+ than does the bulk eluent ion A^-, then the concentration of Q^+ in the adsorbed layer will be higher in the eluite zone than in the neighboring eluent zone. The opposite is true if the eluite ion is less tightly held than the eluent ion. Barber and Carr[53] have applied this in an interesting way to the separation and detection of UV-transparent anions. They employed a UV-absorbing Q^+ species such as alpha-naphthylmethyltripropyl ammonium ion in the mobile phase and with a UV detector monitored the changes in concentration of this species that accompanied the elution of the sample ions. In this way they were able to visualize UV-transparent eluite ions such as chloride and sulfate.

4.5.2c. The Effect of the Modifier

The addition of water miscible organic solvents to the continuous phase of an IIC system reduces the sorption of the IIR. This is a straightforward surface chemical effect of the increasingly nonpolar mobile phase providing a more hospitable environment for the lipophilic IIR. In practical terms it is a very powerful way of controlling the "capacity" of the double layer and in turn the retention of analyte ions.

The retention of small surface inactive eluites in an IIC system where the IIR is lipophilic may therefore be summarized as follows. The lipophilic IIR adsorbs on the surface of the nonpolar stationary phase carrying with it a counterbalancing load of counterions. Eluite ions injected into the mobile phase exchange with the ions of this counterion layer. At the same time competition with counterions from the mobile phase advances the eluite ions through the column but at rates that depend on the relative affinities of the eluite ions for the IIR. In these respects the IIC system is like an ion exchange system. It differs from an ion exchange system in that the capacity of the double layer, which is analogous to the capacity of an ion exchanger, is not a fixed quantity but depends on the composition of the mobile phase and even locally on the presence of the eluite. Theories of the electrical double layer have been applied successfully in a limited number of cases to account for eluite retention.

4.5.3. Retention of Surface Active Eluites

When the analyte species are themselves surface active, the IIR is usually a small surface inactive species such as chloride, perchlorate, or ammonium. The ion exchange model is unsuitable in this case for it is not in the nature of these ions to adsorb significantly at the interface, at least not to the extent of a lipophilic organic ion. Separation of lipophilic analytes can be attributed to selective partitioning between the mobile phase and the interface. Partitioning, in turn, is affected by the hydrophobicity of the analytes, by their tendency to form ion pairs at the interface, and by the various other surface chemical effects that have been discussed above.

4.5.4. IIC and IEC—A Comparison

In most instances where ion exchange chromatography has been applied to solve a problem, ion interaction chromatography has been examined as an alternative. For example, Cassidy and Elchuk[47] gave a detailed description of the separation of inorganic cations and anions on dynamically coated columns. They used alkyl sulfates and sulfonates for cation separations and a variety of quaternary ammonium ions for anions. For stationary phases in this work they used C-18 silica; in later work they used macroporous polystyrene.[54] Separations were good and column efficiencies were high. Figure 4.22 shows a typical separation. Skelly has reported a similar study with emphasis on the determination of organic ions.[55]

Some earlier workers on IIC have cited the relative inefficiency of ion

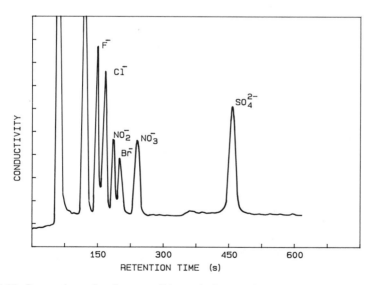

FIG. 4.22. Separation of anions on "dynamic ion exchanger." Experimental conditions: stationary phase was a macroporous polymer (PRP-1); eluent, 7.5×10^{-4} M tetra-butyl ammonium salicylate in acetonitrile–water (7:93). Conductometric detection. (From Ref. 54 with permission.)

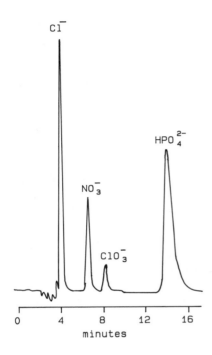

FIG. 4.23. The separation of nitrate and chlorate using ion pair chromatography. Stationary phase was a macroporous polymer (Dionex MPIC-NS1); eluent, 2×10^{-3} M tetrabutylammonium hydroxide in acetonitrile–water (9.5 : 91.5). Suppressed conductometric detection. (From Ref. 58 with permission.)

exchange systems as a major reason for seeking alternatives.[56,57] Nowadays, however, with the availability of ion exchange columns of high efficiencies, the choice of IIC over IEC must be based on other advantages. For the chromatography of small inorganic and organic ions, IEC is usually the more effective and more convenient of the two methods. The need for IIR in the eluent of IIC does add an element of complexity and certainly one of expense, and it can complicate some detection methods.

Lack of selectivity in the ion exchange system can be grounds for preferring the IIC mode. Weiss[58] cites the separation of chlorate and nitrate as such a case. These two ions show very little difference in their selectivities for a typical ion exchanger but can be well separated by IIC (Figure 4.23).

When ions have a very high affinity for an ion exchanger, IIC may be preferred over ion exchange. In ion exchange, to elute tightly held ions within a reasonable time, the only recourse is often a much higher eluent strength. This can create serious problems for some detection methods. With IIC, on the other hand, the extra degree of freedom afforded by the use of organic modifiers to modulate "capacity" is a powerful means for reducing k-primes to practical levels.

The size of the analyte ions can be a deciding factor in favor of IIC. Ion exchange resins as a rule have tight networks that impede ion transfer. If the

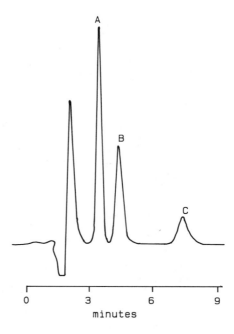

FIG. 4.24. The separation of two closely similar hydroxyalkane sulfonic acids, B and C, by mobile phase ion chromatography. Peaks A, B, and C are due, respectively, to, C_{14}–alkanesulfonate (2-hydroxy), C_{16}–alkanesulfonate(3-hydroxy), and C_{16}–alkanesulfonate(2-hydroxy). Note that B and C differ only in the position of the hydroxy substituent on the alkyl chain. Stationary phase, a macroporous polymer (Dionex MPIC-NS1); eluent, 10^{-2} M ammonium hydroxide in acetonitrile–water (37.5 : 62.5). Suppressed conductometric detection. (From Ref. 58 with permission.)

analyte ions are sufficiently large, their slow mass transfer can lead to poor efficiency with its concomitant penalties of poor resolution and sensitivity. In addition, large organic ions, particularly those with a number of aromatic rings in their structure, have very high affinities for typical styrene-based ion exchangers. In an IIC packing, where the sorption sites are accessible by way of very large pores, there is low mass transfer resistance even for very large ions. This feature has been appreciated for some time; in fact, the application of IIC to large ions predates modern ion chromatography.[45,59]

For the analysis of surface active ions, IIC is invariably the preferred method. Not only are surfactants large and have a high affinity for ion exchangers, but IIC can exploit differences in their hydrophobicity that IEC cannot. Sometimes these differences can be quite subtle. In the numerous examples cited by Weiss[58] the excellent separation of surfactants that differ only in the position of a hydroxy group (Figure 4.24) is a good example of this capability of IIC.

REFERENCES

1. O. D. Bonner and L. L. Smith, A Selectivity Scale for Some Bivalent Cations on Dowex-50, *J. Phys. Chem.* **61,** 326–329 (1957).

2. H. Small, T. S. Stevens, and W. C. Bauman, Novel Ion Exchange Chromatographic Method Using Conductimetric Detection, *Anal. Chem.* **47**, 1801–1809 (1975).

3. D. Reichenberg, Ion Exchange Selectivity, in *Ion Exchange,* Vol. 1 (J. A. Marinsky, ed.), Marcel Dekker, New York (1966).

4. S. Petersen, Anion Exchange Processes, *Ann. N.Y. Acad. Sci.* **57**(Art. 3)**,** 144–158 (1953).

5. F. G. Donnan, Theorie der Membrangleichgewichte und Membranpotentiale bei Vorhandensein von Nicht Dialysierenden Electrolyten. Ein Beitrag zur Physikalisch-chemischen Physiologie, *Z. Elektrochem.* **117**, 572–574 (1911).

6. F. Helfferich, *Ion Exchange,* McGraw-Hill, New York (1962).

7. F. Helfferich, private communication.

8. D. T. Gjerde, G. Schmuckler, and J. S. Fritz, Anion Chromatography with Low-Conductivity Eluents, *J. Chromatogr.* **187**, 35–45 (1980).

9. H. Small and T. E. Miller Jr., Indirect Photometric Chromatography, *Anal. Chem.* **54**, 462–469 (1982).

10. A. Ringbom, *Complexation in Analytical Chemistry,* Interscience Publishers, New York (1963).

11. H. Small, unpublished results.

12. K. A. Kraus and F. Nelson, Anion Exchange Studies of the Fission Products, *Proc. 1st. U.N. Int. Conf. Peaceful Uses Atomic Energy, Geneva,* Vol. 7, pp. 113–125, U.S. Government Printing Office, Washington, D.C. (1956).

13. J. S. Fritz and D. J. Pietrzyk, Non-aqueous Solvents in Anion-Exchange Separations, *Talanta* **8**, 143–162 (1961).

14. I. Hazan and J. Korkisch, Anion-Exchange Separation of Iron, Cobalt, and Nickel, *Anal. Chim. Acta* **32**, 46–51 (1965).

15. F. W. E. Strelow, C. R. van Zyl, and C. J. C. Bothma, Distribution Coefficients and Cation Exchange Behavior of Eluents in Hydrochloric Acid–Ethanol Mixtures, *Anal. Chim. Acta* **45**, 81–92 (1969).

16. J. S. Fritz, D. T. Gjerde, and C. Pohlandt, *Ion Chromatography,* Dr. Alfred Huthig Verlag, Heidelberg (1982).

17. Y. Marcus and A. S. Kertes, *Ion Exchange and Solvent Extraction of Metal Complexes,* Wiley-Interscience, New York (1969).

18. A. E. Martell and R. M. Smith, *Critical Stability Constants,* Vol. 3, Plenum Press, New York (1977).

19. H. Small, unpublished results.

20. J. M. Riviello and C. A. Pohl, Ion Chromatography of Transition Metals, presented at the 25th Rocky Mountain Conference, Denver, August 14–17 (1983).

21. J. M. Riviello and C. A. Pohl, Advances in the Separation and Detection of Transition and Post-transition Metals by Ion Chromatography, paper No. 70, 35th Pittsburgh Conference on Analytical Chemistry, Atlantic City, March 1984.

22. F. G. Helfferich, "Ligand Exchange": A Novel Separation Technique, *Nature* **189**, 1001–1002 (1961).

23. V. A. Davankov and A. V. Semechkin, Ligand-Exchange Chromatography, *J. Chromatogr.* **141**, 313–353 (1977).

24. H. Small, Porous Cation Exchange Resins as Selective Sorbents in Organic Systems, U.S. Patent No. 3,409,691 (1968).

25. H. P. Gregor, A General Thermodynamic Theory of Ion Exchange Processes, *J. Am. Chem. Soc.* **70**, 1293 (1948).

26. H. P. Gregor, Gibbs–Donnan Equilibria in Ion Exchange Resin Systems, *J. Am. Chem. Soc.* **73**, 642–650 (1951).

27. J. I. Bregman, Cation Exchange Processes. *Ann. N.Y. Acad. Sci.* **57** (Art. 3)**,** 125–143 (1953).

28. G. Eisenmann, The Electrochemistry of Cation-Sensitive Glass Electrodes, in *Advances in*

Analytical Chemistry and Instrumentation, Vol. 4 (C. N. Reilley, ed.), Wiley-Interscience, New York (1965).

29. R. M. Diamond and D. C. Whitney, Resin Selectivity in Dilute to Concentrated Aqueous Solutions, in *Ion Exchange,* Vol. 1 (J. A. Marinsky, ed.), Marcel Dekker, New York (1966).

30. W. F. McDevit and F. A. Long, The Activity Coefficient of Benzene in Aqueous Salt Solutions, *J. Am. Chem. Soc.* **74,** 1773–1777 (1952).

31. G. L. Schmitt and D. J. Pietrzyk, Liquid Chromatographic Separation of Inorganic Anions on an Alumina Column, *Anal. Chem.* **57,** 2247–2253 (1985).

32. R. G. Pearson, Hard and Soft Acids and Bases, in *Collected Readings in Inorganic Chemistry,* Vol. 2, pp. 141–152 (G. Galloway, ed.), Chemical Education Publishing Co., Easton, Pennsylvania (1972).

33. R. W. Slingsby and C. A. Pohl, Anion-Exchange Selectivity in Latex-Based Columns for Ion Chromatography, *J. Chromatogr.* **458,** 241–253 (1988).

34. J. M. Riviello, Complex Equilibria and the Analytical Ion-Exchange Properties of Divalent transition metals, in *Ion Exchange Technology* (D. Naden and M. Streat, eds.), pp. 585–594, Ellis Horwood Limited, West Sussex, England (1984). Distributed by John Wiley and Sons, ISBN 0-85312-770-0.

35. G. W. Boyd, A. W. Adamson, and L. S. Myers Jr., The Exchange Adsorption of Ions from Aqueous Solutions of Organic Zeolites. II. Kinetics, *J. Am. Chem. Soc.* **69,** 2836–2849 (1947).

36. B. A. Soldano, The Kinetics of Ion Exchange Processes, *Ann. N.Y. Acad. Sci.* **57**(Art.3), 116–124 (1953).

37. E. Glueckauf, Theory of Chromatography. Part 9. The "Theoretical Plate" Concept in Column Separations, *Trans. Faraday Soc.* **51**(385), Pt. 1, 34–44 (1955).

38. E. Glueckauf, Principles of Operation of Ion-Exchange Columns, in *Ion Exchange and its Applications,* pp. 27–38, Society of Chemical Industry, London (1955).

39. L. C. Hansen and T. W. Gilbert, Theoretical Design for High-Speed Liquid and Ion Exchange Chromatography, *J. Chromatogr. Sci.* **12**(8), 464–472 (1974).

40. T. S. Stevens and H. Small, Surface Sulfonated Styrene Divinyl Benzene—Optimization of Performance in Ion Chromatography, *J. Liq. Chromatogr.* **1**(2), 123–132 (1978).

41. J. C. Giddings, The Role of Lateral Diffusion as a Rate-Controlling Mechanism in Chromatography, *J. Chromatogr.* **5,** 46–60 (1961).

42. J. C. Giddings, Theory of Chromatography, in *Chromatography* (E. Heftmann, ed.), pp. 20–32, Reinhold, New York (1961).

43. Reference 6, pp. 467–468.

44. F. H. Spedding, J. E. Powell, and E. J. Wheelwright, Separation of Adjacent Rare Earths with Ethylenediamine–Tetraacetic Acid by Elution from an Ion-Exchange Resin, *J. Am. Chem. Soc.* **76,** 612–613 (1954).

45. L. R. Snyder and J. J. Kirkland, *Introduction to Modern Liquid Chromatography,* John Wiley and Sons, New York (1979).

46. Z. Iskandarani and J. Pietrzyk, Ion Interaction Chromatography of Organic Anions on a Poly(styrene-divinylbenzene) Adsorbent in the Presence of Tetraalkylammonium Salts, *Anal. Chem.* **54,** 1065–1071 (1982).

47. R. M. Cassidy and S. Elchuk, Dynamically Coated Columns for the Separation of Metal Ions and Anions by Ion Chromatography, *Anal. Chem.* **54,** 1558-1563 (1982).

48. C. Horvath, W. Melander, I. Molnar, and P. Molnar, Enhancement of Retention by Ion-Pair Formation in Liquid Chromatography with Non-polar Stationary Phases, *Anal. Chem.* **49,** 2295–2305 (1977).

49. B. A. Bidlingmeyer, S. N. Deming, W. P. Price, B. Sachok, and M. Petrusek, Retention Mechanisms for Reversed-Phase Ion-Pair Liquid Chromatography, *J. Chromatogr.* **186,** 419–434 (1979).

50. A. W. Adamson, *Physical Chemistry of Surfaces,* 3rd edition. John Wiley and Sons, New York (1976).
51. F. F. Cantwell and S. Puon, Mechanism of Chromatographic Retention of Organic Ions on a Nonionic Adsorbent, *Anal. Chem.* **51,** 623–632 (1979).
52. L. Hackzell and G. Schill, Detection by Ion-Pairing Probes in Reversed-Phase Liquid Chromatography, *Chromatographia* **15,** 437–444 (1982).
53. W. E. Barber and P. W. Carr, Ultraviolet Visualization of Inorganic Ions by Reversed-Phase Ion-Interaction Chromatography, *J. Chromatogr.* **260,** 89–96 (1983).
54. R. M. Cassidy and S. Elchuk, Dynamically Coated Columns and Conductivity Detection for Trace Determination of Organic Anions by Ion Chromatography, *J. Chromatogr.* **262,** 311–315 (1983).
55. N. E. Skelly, Separation of Inorganic and Organic Anions on Reversed-Phase Liquid Chromatography Columns, *Anal. Chem.* **54,** 712–715 (1982).
56. I. Molnar, H. Knauer, and D. Wilk, High-Performance Liquid Chromatography of Ions, *J. Chromatogr.* **201,** 225–240 (1980).
57. R. N. Reeve, Determination of Inorganic Main Group Anions by High-Performance Liquid Chromatography, *J. Chromatogr.* **177,** 393–397 (1979).
58. J. Weiss, *Handbook of Ion Chromatography,* Dionex Corporation, Sunnyvale, California (1986).
59. J. N. Done, J. H. Knox, and J. Loheac, *Applications of High-Speed Liquid Chromatography,* John Wiley and Sons, New York (1974).

Chapter 5

Ion Exchange Resins in Liquid Partition Chromatography

5.1. INTRODUCTION

There are a number of important separation techniques that use ion exchange resins but in ways that do not involve the ion exchange reactions of the resin. In modern IC, ion exclusion is perhaps the best known of these techniques as it is widely used for the separation of organic acids. In this application a strong acid cation exchange resin in the hydrogen form is the stationary phase and usually a dilute mineral acid solution is the mobile phase. Since anionic species are being separated on a cation exchanger it is clear that some process other than ion exchange is involved. It is generally accepted that the mechanism is one of selective partitioning between the resin phase and the external aqueous phase. The degree to which a solute partitions into the resin phase depends on its solubility in that phase, and when the solute is ionic or has partial ionic character the solubility can be strongly modified by electrostatic forces between the phases. One purpose of this chapter is to examine the several factors that affect the partitioning process and show how they may be used to control and improve separations.

5.2. SORPTION EQUILIBRIA ON ION EXCHANGE RESINS

Discussion of sorption or solubility phenomena is subdivided as follows:

1. Sorption of uncharged species into ion exchange resins.
2. Sorption of ionic materials.
3. Sorption of weak electrolytes.

5.2.1. Sorption of Uncharged Species

When ion exchange resins are contacted with aqueous solutions of uncharged molecules, sorption of the uncharged species is observed. Wheaton and Bauman[1,2] were among the first to study this phenomenon and to show how it might be used in the separation of water-soluble nonionic materials. Table 5.1 lists the distribution coefficients determined by Wheaton and Bauman for a variety of molecules on typical ion exchange resins. Reichenberg and Wall [3] made a systematic study of the distribution of uncharged molecules between a resin and solution phase in which they varied the degree of cross-linking of the resin. They concluded that the distribution could be treated as a dissolution in the resin phase which is modified in an intricate way by such additional factors as salting out by the resin, London dispersion (van der Waals) interactions of the

TABLE 5.1. Distribution Coefficients (K_D) for the Partitioning of Solutes between Water and Various Ion Exchange Resins[a]

Solute	Resin	K_D
Ethylene glycol	Dowex 50 ×8, H$^+$	0.67
Sucrose	Dowex 50 ×8, H$^+$	0.24
d-Glucose	Dowex 50 ×8, H$^+$	0.22
Glycerine	Dowex 50 ×8, H$^+$	0.49
Triethylene glycol	Dowex 50 ×8, H$^+$	0.74
Phenol	Dowex 50 ×8, H$^+$	3.08
Acetic acid	Dowex 50 ×8, H$^+$	0.71
Acetone	Dowex 50 ×8, H$^+$	1.2
Formaldehyde	Dowex 50 ×8, H$^+$	0.59
Methyl alcohol	Dowex 50 ×8, H$^+$	0.61
Formaldehyde	Dowex 1 ×7.5, Cl$^-$	1.06
Acetone	Dowex 1 ×7.5, Cl$^-$	1.08
Glycerine	Dowex 1 ×7.5, Cl$^-$	1.12
Methyl alcohol	Dowex 1 ×7.5, Cl$^-$	0.61
Phenol	Dowex 1 ×7.5, Cl$^-$	17.7
Xylose	Dowex 50 ×8, Na$^+$	0.45
Glycerine	Dowex 50 ×8, Na$^+$	0.56
Pentaerythritol	Dowex 50 ×8, Na$^+$	0.39
Ethylene glycol	Dowex 50 ×8, Na$^+$	0.63
Diethylene glycol	Dowex 50 ×8, Na$^+$	0.67
Triethylene glycol	Dowex 50 ×8, Na$^+$	0.61
Ethylene diamine	Dowex 50 ×8, Na$^+$	0.57
Diethylene triamine	Dowex 50 ×8, Na$^+$	0.57
Triethylene tetramine	Dowex 50 ×8, Na$^+$	0.64
Tetraethylene pentamine	Dowex 50 ×8, Na$^+$	0.66

[a]Reference 1.

solute with the hydrocarbon matrix, and interactions with the counterion on the resin.

Although there is as yet no quantitative theory for the distribution in these systems, there are satisfactory qualitative explanations for most of the effects. Thus, factors that oppose sorption, such as salting out, may be expected to increase with increasing cross-linkage of the resin since this results in a higher effective "ionic strength" in the resin phase. On the other hand, London dispersion interactions which promote sorption will be more significant as the matrix/ water ratio of a resin increases, as it does the higher the cross-linkage of the resin.

The structure of the solute itself has a major influence on its distribution. Molecules with nonpolar substituents tend to favor the resin phase over those that are more polar. Aromatic substituents in the solute enhance sorption onto polystyrene-based resins. Within an homologous series there is an increase in sorption with chain length up to the point where steric exclusion by the resin network opposes the intrusion of larger molecules. Furthermore, there is often a regularity in this progression that is the same as that observed in other forms of chromatography and separation systems in that log K_D is proportional to the number of carbons in the nonpolar part of the molecule. Langmuir observed the effect in solid phase adsorption and ascribed it to a "principle of independent surface action." Martin[4] noted a similar dependence in partition chromatography, which he explained by assuming that the interaction energy of a molecule could be approximated by a summation of individual group energies. Thus in transferring an alcohol molecule from one phase to another the total free energy change associated with the transfer, ΔF, could be considered as the sum of the free energy changes of the separate groups, that is,

$$\Delta F = n\Delta F_{-CH_2} + \Delta F_{-OH} \qquad (5.1)$$

since

$$\Delta F = RT \ln K \qquad (5.2)$$

then

$$\ln K \propto n\Delta F_{-CH_2} + \Delta F_{-OH} \qquad (5.3)$$

where K is the distribution ratio. Since the end group is fixed in a homologous series and ΔF_{-OH} is therefore constant the reason for the linearity between log K and carbon number is clear. Small and Bremer [5] observed this relationship to apply to the distribution of alcohols between water and "amphiphilic resins."

Since London or hydrophobic interactions promote sorption while salting out by the resin opposes it, a number of workers have sought to increase the former relative to the latter. Sargent and Rieman[6,7] found that adding electrolyte to the external phase greatly increased the distribution coefficient in favor of the resin. Adding electrolyte tends to equalize the salting out effect of both phases and allows London forces to dominate. Ammonium sulfate was the electrolyte most used by them in their studies, and Figure 5.1 is illustrative of the type of behavior they observed. These workers developed many separation schemes using solutions of ammonium sulfate as eluents. They called this technique "salting-out chromatography" (SOC).

Although SOC has been used principally for the separation of nonionic species, in principle it ought to be successful for ionic species as well. Adding a large amount of electrolyte to the external phase will reduce the Donnan potential (see Section 4.2.2) and allow species to enter the resin that are normally excluded. Then hydrophobic interactions between the ions and the resin matrix will tend to determine the distribution and some useful separations should be possible. Detection by photometric or electrochemical methods is likely to be the most suitable in view of the high concentration of electrolyte present.

Sargent and Graham[8] studied a series of low-capacity resins as sorbents for organic acids and amines. Their motivation was to prepare resins for large-scale scavenging of trace organics from water that could compete with charcoal on the sorption part of the cycle but be more readily regenerated. They prepared a series of partially but uniformly sulfonated resins ranging in capacity from 1.37 to 5.29 meq per dry gram—the latter being a fully sulfonated resin. These materials

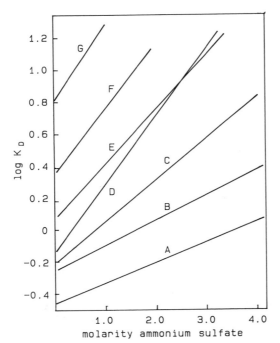

FIG. 5.1. Distribution of alcohols between Dowex-1 and ammonium sulfate. Distribution coefficients as a function of ammonium sulfate concentration. A, glycerol; B, methyl alcohol; C, ethyl alcohol; D, t-butyl alcohol; E, n-propyl alcohol; F, n-butyl alcohol; G, n-amyl alcohol. (From Ref. 6 with permission. Copyright 1957 American Chemical Society.)

TABLE 5.2. Distribution Coefficients (K_D) for the Partitioning of Acids between Partially Sulfonated Polystyrene Divinylbenzene Polymers and Water[a]

Acid sorbed	Ionic form	Resin properties			
		Capacity meq/g (dry H^+ form)	Water content (%)	Percent DVB	K_D
Phenol	Na^+	5.29	73	2	1.02
Phenol	Na^+	2.73	60	2	24.2
Phenol	Na^+	2.17	58	2	29.6
Phenol	Na^+	1.58	47	2	65.0
Phenol	Na^+	1.06	38	2	56.7
Phenol	H^+	1.37	48	2	24.9
Propionic	H^+	1.37	48	2	3.29
Valeric	H^+	1.37	48	2	18
Caprylic	H^+	1.37	48	2	357
Malonic	H^+	1.37	48	2	0.89
Succinic	H^+	1.37	48	2	1.16
Adipic	H^+	1.37	48	2	2.34
Benzoic	H^+	1.37	48	2	37.8

[a]Reference 8. Copyright 1962 American Chemical Society.

showed enhanced sorption for organic acids and phenol as capacity was lowered (Table 5.2). Lowering the capacity of the resin lowered its salting out tendency while at the same time increasing its hydrophobicity. Both effects thus acted in a cooperative way and improved the resin's affinity for nonpolar organics.

Regeneration with sodium hydroxide also proved to be very effective as exemplified by the case of phenol extraction (Figure 5.2). Donnan exclusion forces effectively expelled the phenate ion from the interior of the cation exchanger and permitted phenol to be recovered in a very small volume relative to the amount of feed that had been "purified."

Although this work was directed towards large-scale applications, it has implications for analysis. To the author's knowledge this type of resin has not been used in IC, but the concentration capabilities alone suggest a utility in the concentration of trace organic acids from aqueous streams as a preliminary to their analysis.

Small and Bremer[5] found that the hydrophobicity of an ion exchanger could be markedly altered by the simple expedient of replacing the usual small hydrophilic counterion (e.g., Cl, Na) with a large amphiphilic counterion. Most of their work involved the modification of common anion exchangers such as Dowex 1 with counterions such as di-2-ethyl hexyl phosphate or stearate. They envisioned the resin as comprising hydrophilic and lipophilic regions as a result

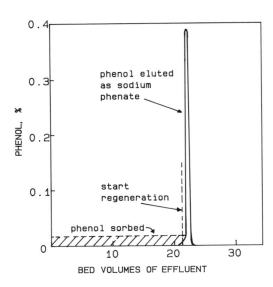

FIG. 5.2. Sorption and elution of phenol on a low-capacity cation exchanger in the Na$^+$ form. Phenol at 100 ppm was loaded for about 22 bed volumes onto the low-capacity resin. This was followed by a small volume of 1.0 N NaOH that effectively eluted the bulk of the sorbed phenol in about one bed volume. (From Ref. 8 with permission. Copyright 1962 American Chemical Society.)

of pseudomicelles forming within the resin phase. Resins of this type showed significantly enhanced partitioning of alcohols when compared with a more conventional resin form and displayed some unusual thermal effects that were exploited to enhance separations of various mixtures.

In summary, ion exchange resins are good sorbents for nonelectrolytes. Dispersion-type interactions between the resin matrix and the solute promote sorption while salting out by the resin "electrolyte" opposes it. Sorption may be significantly increased by increasing the hydrophobicity of the resin or by adding electrolyte to the external phase. However, most IC applications that employ resins in a partitioning mode use conventional resins in conjunction with water or dilute (< 0.1 M) electrolytes.

5.2.2. Sorption of Strong Electrolytes

Electrolytes, in contrast to nonelectrolytes, are poorly sorbed by ion exchange resins. To illustrate the physical basis for this low solubility of electrolytes we will consider the distribution of hydrochloric acid between water and a strong acid-type cation exchange resin in the hydrogen form.

We saw in Section 4.2.2 that when an ion exchange resin is placed in contact with water, a potential develops between the two phases. This is the Donnan potential (DP). In the case of a cation exchanger the internal phase becomes negative with respect to the external phase while just the opposite is the case for an anion exchanger. Section 4.2.2 treated the distribution of counterions between the two phases and how the DP affected it; in this section we will be concerned primarily with the distribution of the co-ions.

In our model case the co-ion is the chloride ion, which "sees" a strong negative potential resisting its intrusion into the resin phase, so at equilibrium the concentration of chloride in the resin tends to be much lower than in the external phase. The chloride ion must be accompanied by an equivalent amount of hydronium ion to preserve electroneutrality, so the net effect of the distribution is that the resin phase contains hydrochloric acid but at a much lower concentration than that in the external phase. Now as the acid level in the external phase is raised, the DP diminishes since the escaping tendencies of the hydronium ions in the two phases approach one another. This in turn reduces the rejection of chloride and the resin becomes increasingly invaded by hydrochloric acid. This is a prominent feature of electrolyte sorption, namely, that it increases in a disproportionate manner with increasing electrolyte concentration, or as will be seen presently, the distribution coefficient increases with increasing concentration.

Other effects can be explained on the basis of the Donnan potential. Thus electrolyte exclusion is greater the higher the cross-linkage of the resin since the higher effective molarity of the higher cross-linked resins means a higher escaping tendency of counterions from the resin phase, hence a greater DP.

The higher the valence of the co-ion the more it is excluded since it experiences a greater repulsive force because of its higher charge. Thus sodium sulfate will be more excluded than sodium chloride from a sodium form resin.

5.2.3. Thermodynamic Treatment of Electrolyte Distribution and Ion Exclusion

At equilibrium the chemical potentials of all mobile ionic species will be the same in both phases. The following equations express the chemical potentials of hydronium and chloride ions in the model system of hydrogen form resin in contact with hydrochloric acid. In treating the Donnan potential the resin phase is arbitrarily chosen to be at zero potential so the solution phase is at the DP, which is denoted as ψ. Thus for the hydronium ion in the two phases

$$(\mu H)_R = \mu_0^* + RT \ln[H^+]_R \qquad (5.4)$$

$$(\mu H)_S = \mu_0^* + RT \ln[H^+]_S + \psi Z_H F \qquad (5.5)$$

where μ_0^* is the chemical potential of the hydronium ion at some standard state. The third term on the right-hand side of (5.5) represents the electrostatic contribution to the total electrochemical potential of the species.

Concentrations have been used in place of activities but the approximation does not invalidate the main conclusions to be drawn from this simple treatment.

For chloride ion in the two phases

$$(\mu_{cl})_R = \mu_0 + RT \ln[Cl^-]_R \qquad (5.6)$$

$$(\mu_{Cl})_R = \mu_0 + RT \ln[Cl^-]_S - \psi Z_{Cl}F \qquad (5.7)$$

where μ_0 is the chemical potential of chloride at some standard state. Note that the electrostatic term is opposite in sign to that of (5.5) since the charge of the chloride, Z_{Cl}, is opposite to that of the hydronium ion.

At equilibrium, chemical potentials of species are the same in both phases so equating (5.4) and (5.5) and (5.6) and (5.7) gives

$$\psi = (RT/Z_HF)\ln([H^+]_R/[H^+]_S) \qquad (5.8)$$

$$\psi = -(RT/Z_{Cl}F)\ln([Cl^-]_R/[Cl^-]_S) \qquad (5.9)$$

In turn, eliminating ψ from (5.8) and (5.9) gives

$$\ln([Cl^-]_R/[Cl^-]_S) = \ln([H^+]_S/[H^+]_R) \qquad (5.10)$$

or simply

$$[Cl^-]_R/[Cl^-]_S = [H^+]_S/[H^+]_R \qquad (5.11)$$

The hydronium ion in the resin is made up of two contributions, the hydronium counterion concentration, which equals the capacity of the resin, C_R, and the amount that accompanies the chloride in the resin, so $[H^+]_R = C_R + [Cl^-]_R$. Thus (5.11) becomes

$$[Cl^-]_R/[Cl^-]_S = [H^+]_S/(C_R + [Cl^-]_R) \qquad (5.12)$$

In dilute solutions $[Cl^-]_R$ will be much less than C_R so

$$[Cl^-]_R/[Cl^-]_S = [Cl^-]_S/C_R \qquad (5.13)$$

In higher solution concentrations, of course, the approximation is invalid, but since IC is concerned mainly with exclusion from dilute solutions the approximation is permissible.

The term on the left-hand side of (5.13) is the distribution coefficient for chloride, or HCl for that matter, in the resin–solution system. For low external concentrations of acid and conventional resins where C_R is approximately 3 meq/ml it is clear that the distribution coefficient for the HCl is very low. Thus if dilute HCl is injected into a chromatographic system of a hydrogen form resin, and water is the eluent, it will pass through the column essentially unretained.

This difference in distribution behavior between electrolytes and non-electrolytes is the basis of ion exclusion, a separation method first proposed by Wheaton and Bauman.[2] They targeted ion exclusion mainly at large-scale industrial applications such as the removal of contaminating electrolytes from water soluble organics. The removal of salt from glycerol or glycols was one of their well-studied examples. Today the ion exclusion principle is widely employed in IC, particularly in the analysis of mixtures of organic acids. To understand the factors involved we need to examine the distribution of *weak* electrolytes between a resin and an external aqueous phase.

5.2.4. Sorption of Weak Electrolytes

The partitioning of weak electrolytes may best be illustrated by considering the partitioning of organic acids. Not only do they embody a number of important features that determine sorption, but their analysis by ion exclusion is an important part of IC methodology. Initially we will consider the distribution of an acid, denoted HA, between water and a hydrogen form of a sulfonated resin. Because the acid is partially ionized in water

$$HA \rightleftharpoons H^+ + A^- \tag{5.14}$$

the partitioning into the resin depends on its degree of ionization. Neutral HA molecules are unaffected by the Donnan potential and, barring any steric restrictions, can freely enter the resin. Anionic species A^- are, on the other hand, essentially excluded from the resin phase. So the stronger the acid the more it tends to be rejected by the resin. On the other hand, dispersive interactions between HA and the resin matrix oppose this and promote sorption.

The several distribution equilibria may be described as follows. The overall distribution coefficient, K_D, for all species involving A (i.e., A^- and HA) is the ratio of the concentration of A species in the resin phase to the concentration of A species in the external solution phase, that is,

$$K_D = ([HA]_R + [A^-]_R)/([HA]_S + [A^-]_S) \tag{5.15}$$

This is the distribution coefficient that determines the migration rate of a band of acid in a chromatographic operation. The overall distribution is determined in turn by the two separate but related equilibria, the distribution of neutral HA molecules and of the ionic species A^-. The distribution of HA can be described by the expression

$$k_{HA} = [HA]_R/[HA]_S \tag{5.16}$$

Similarly for A^-

$$k_{A^-} = [A^-]_R/[A^-]_S \tag{5.17}$$

where k_{HA} and k_{A^-} are the distribution coefficients for the subscripted species. Substituting in (5.15) gives

$$K_D = (k_{HA}[HA]_S + k_{A^-}[A^-]S)/([HA]_S + [A^-]_S) \tag{5.18}$$

Because of Donnan exclusion, k_{A^-} is very small, so

$$K_D = k_{HA}[HA]_S/([HA]_S + [A^-]_S) \tag{5.19}$$

Now $[HA]_S/([HA]_S + [A^-]_S)$ is simply the degree of association of the acid, usually expressed as $(1 - \alpha)$, where α is the fraction of acid dissociated, so (5.19) becomes

$$K_D = k_{HA}(1 - \alpha) \tag{5.20}$$

This very simple expression embodies the two major factors that control the distribution of weak acids in the system. Thus k_{HA} represents the sorption-promoting tendency while $(1 - \alpha)$ represents the ion exclusion effect.

Since the degree of association of a weak acid decreases with increasing dilution so also does K_D decrease. In chromatography this leads to increasingly less retention as the sample is diluted. It also means that with water as eluent, peaks will tend to exhibit fronting (see Section 2.3.5). This is the opposite of the more common type of peak skewing that occurs when K_D's rise with decreasing solute concentration.

One can get an estimate of the magnitude of K_D by examining curves for acid dissociation equilibria. Figure 5.3 is a plot of the fraction of acid associated as a function of acid concentration. The curves represent acids of different pK_a. In IC, acids are usually injected at concentrations of 10^{-3} M or less so it is appropriate to examine the degree of association at or below that concentration. For significant retention of an acid in a chromatographic mode its k-prime should be at least 0.1, which implies a K_D of roughly half that value. From Figure 5.3 it

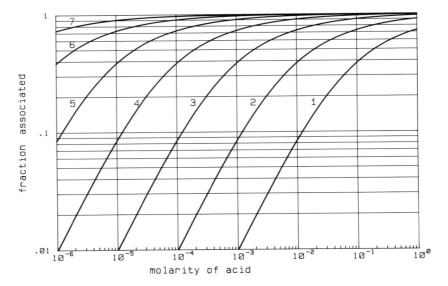

FIG. 5.3. The association of weak acids as a function of acid concentration. Curve 1 represents an acid of $pK = 1$, curve 2 an acid of $pK = 2$, and so on.

is clear that for acids of pK_a one or less, sorption is unlikely. Since the value of $(1 - \alpha)$ is less than 0.01 it would require a large counterbalancing value of k_{HA} to give an adequately large K_D. That is unlikely for small molecules so acids of this strength tend to elute at the void.

As the acid strength falls, the degree of association rises and retention can be expected to increase. For example, an acid of $pK_a = 5$ is 90% associated at 10^{-3} M, so retention is predicted, especially if sorption is augmented by a relatively large k_{HA}.

Although it is possible to separate weak acids on a cation exchange resin using water as eluent it is more common to use a dilute mineral acid or some other strong acid as eluent. There are two major reasons for doing this. In the first place the strong acid increases association of the organic acid; this diminishes the Donnan exclusion effect, thus allowing dispersion forces to become dominant. This can be effective in raising the k-prime of the acid to a useful level. Secondly, the organic acid elutes in an environment of essentially constant pH so there is no variation in K_D across the band. This eliminates one of the major causes of band skewing. A drawback of using acid as an eluent is that it complicates conductometric detection of the eluted acids; there are, however, effective means for coping with the background acid (Chapter 9).

Work by Turkelson and Richards[9] provides good examples of how these various effects are manifested in a practical chromatographic situation. These authors studied the elution of several organic acids from Dowex 50 type resins in the H^+ form, using both water and dilute hydrochloric acid solutions as eluents. Among the many acids they examined were mono- di-, and tricholoroacetic acids.

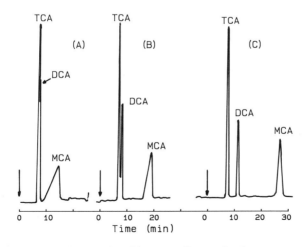

FIG. 5.4. Separation of chloroacetic acids on a sulfonated polystyrene type resin in the H^+ form, using A, water as eluent; B, 10^{-3} M HCl as eluent; C, 10^{-2} M HCl as eluent. (From Ref. 9 with permission. Copyright 1978 American Chemical Society.)

Figure 5.4 shows the chromatograms obtained when water, 0.001 M HCl, and 0.01 M HCl were used as eluents. It is instructive to examine these chromatograms alongside the data of Figures 5.5 and 5.6. Figure 5.5 is a plot of the fraction associated versus organic acid concentration, while Figure 5.6 is a plot of association versus the concentration of HCl in the mobile phase.

With water as eluent, little retention of DCA or TCA is to be expected since they are too highly dissociated. Chromatogram 5.4A bears that out. Using 10^{-3} M HCl does not alter that much, and this is reflected in the very slight improvement in separation of these two components evident in chromatogram B. However, in 10^{-2} M there is now significant association of DCA relative to TCA (Figure 5.6), which correlates with the pulling apart of these two species as seen in chromatogram C.

Monochloracetic acid, the weakest of the three acids, can be expected to exhibit retention in all three media. In water it is significantly associated at 10^{-3} M but becomes increasingly less so as it dilutes. So besides expecting retention, we would also expect band skewing as the dilute downstream side of the band becomes progressively more diffuse. In 10^{-3} M HCl, MCA is more associated than it is in water so greater retention is predictable. Furthermore the degree of association is maintained as the band elutes since the pH remains essentially fixed. With water, on the other hand, the K_D drops as the band dilutes, causing it to speed up as elution proceeds. The retention of MCA is more pronounced in

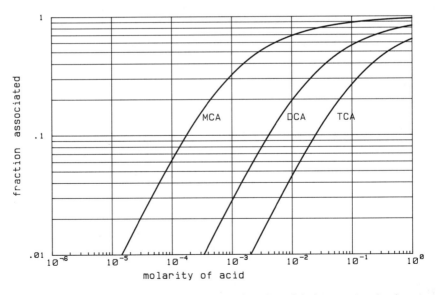

FIG. 5.5. Association of chloroacetic acids as a function of their concentration in water.

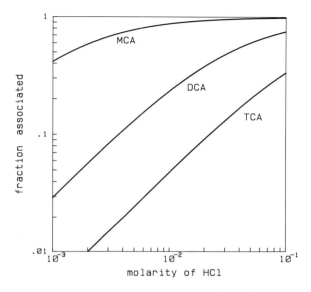

FIG. 5.6. Association of chloroacetic acids ($<10^{-4}$ M) in HCl as a function of HCl concentration.

10^{-2} M HCl; that too correlates well with what one would predict from Figure 5.6.

The shape of the MCA peak also fits well with our partitioning mechanism. Fronting is very evident in chromatogram A but is much less when HCl is present. The uniform pH throughout the organic acid band in the presence of HCl assures that all zones within the band move with the same K_D, a recipe for symmetrical peaks.

Before leaving the topic of ion exclusion, a comment on the terminology may be appropriate. As originally introduced by Wheaton and Bauman, the term "ion exclusion" aptly described the process; in water media the resin phase excluded electrolytes and admitted nonionics. But ion exclusion as used in the context of modern IC is something of a misnomer. In its most commonly practiced form, a strong acid is added to the mobile phase to *reduce* ionization and, as a result, ion exclusion. In the case of acids of borderline strength, Donnan exclusion will certainly exert an important effect on partitioning, but for most weak acids the partitioning process can be described as conventional partitioning of neutral species, the undissociated acids, between the resin and mobile phase. Looked at from this point of view, the term "ion exclusion" seems an inappropriate choice, yet common usage probably ensures that it will continue to be linked with this type of separation. Here is a good example of how terminology can actually be misleading in a sense, and how important it is to be aware of the principles that underlie the term, however appropriately it is applied.

5.3. WATER AS ELUENT IN ION CHROMATOGRAPHY

There are a number of ion chromatographic techniques that, though not widely used, deserve mention because of their simplicity and their potential for future development. All use water or other polar solvents as the sole component of the mobile phase. This not only eliminates the need for electrolytes and precise eluent make-up, but it avoids many of the detection problems that ionic eluents can impose (see Chapter 7).

Since Pedersen[10] prepared the first crown ethers and demonstrated their ability to complex cations there has been considerable interest in applying this discovery in a chromatographic way. Blasius and co-workers have worked very extensively in the area of coupling functionality of this sort into both polymeric and inorganic structures. Their work has been the subject of a number of thorough reviews.[11,12]

Figure 5.7 illustrates examples of the type of groups that have been bonded either in polymers or to silica substrates to form electrolyte sorbents. Both water and methanol have been studied as solvents in the external phase; the extraction is usually stronger out of methanol than it is out of water.

Many chromatographic separations have been attempted using these crown ether sorbents as mobile phase and usually water or methanol as mobile phase.

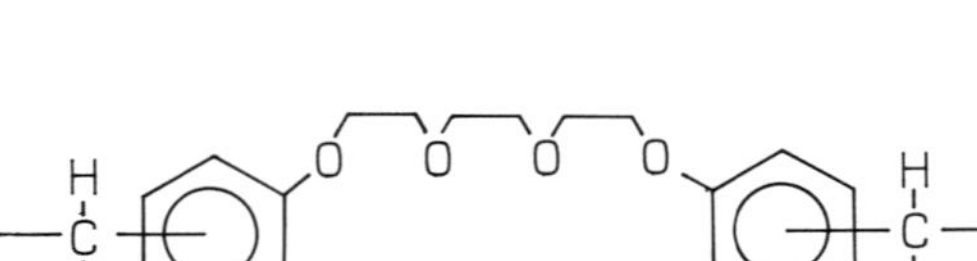

dibenzo-14-crown-4

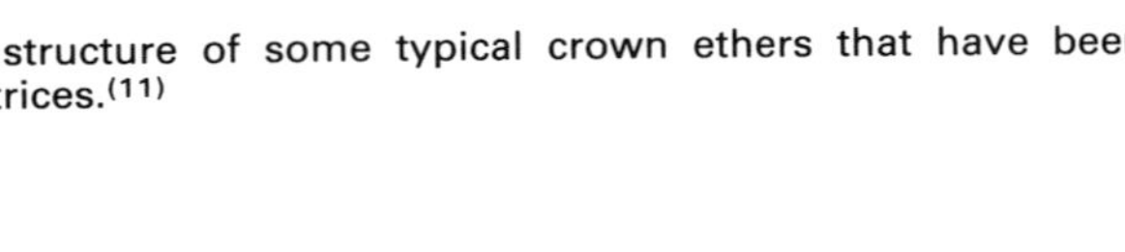

dibenzo-21-crown-7

FIG. 5.7. The structure of some typical crown ethers that have been attached to insoluble matrices.[11]

The earliest attempts, while they gave separations, were characterized by having rather broad peaks even under low-flow-rate conditions. More recently there has been considerable improvement in performance.[13]

Both cations and anions can be separated by these materials. Figures 5.8 and 5.9 show alkali metals separated as their chlorides and several anions as their potassium salts. When the sample contains a mixture of cations as well as anions it is necessary to pretreat the sample to avoid "scrambling" into a confusing variety of ion pairs. Pretreatment involves a simple ion exchange that converts all cations to a common cation, if anion analysis is the objective. A small cation exchange bed inserted between the sample injection valve and the separator column can accomplish this; an anion exchange column if cation analysis is required.

A paper by Kimura *et al.*[13] reports quite efficient separation of alkali metal and alkaline earth ions using silica columns modified *in situ* with lipophilic crown ethers. Clean separations of Li, Na, K, Rb, and Cs are accomplished in less than three minutes.

Hatch and co-workers in 1957 reported a new form of ion separation that they called "ion retardation."[14] Chromatographic in nature, although intended primarily for large-scale separation, it employed a unique type of resin that Hatch imaginatively called a "snake-cage resin." The most studied example comprised a polyacrylate anion "snake" entrapped in an anion exchange resin (Dowex 1) "cage." Since the polyanion and the resin polycation are both immobile, in principle, no Donnan potential can develop and anions and cations should have equal access to the interior of such a resin. Thus ion retardation resins imbibe considerable amounts of electrolytes from even dilute solutions; in fact the co-

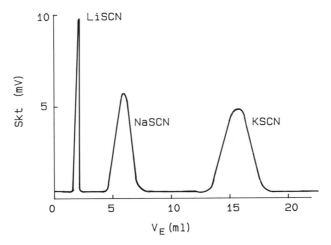

FIG. 5.8. Separation of cations as thiocyanates on a crown-ether resin (dibenzo-21-crown-7). Water eluent; potentiometric detection. (From Ref. 11 with permission.)

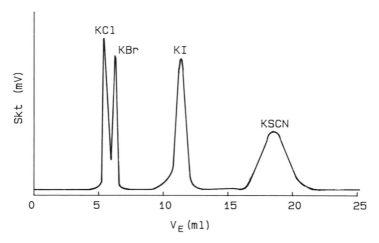

FIG. 5.9. Separation of anions as their potassium salts on an exchanger with dibenzo-21-crown-7 anchor groups. Water eluent. (From Ref. 11 with permission.)

efficient for electrolyte distribution increases with increasing dilution of the external phase. Sorption is also selective with an affinity order for cations and anions paralleling that observed for ion exchange resins containing similar functionalities. The particle size used was large—typically 20–50 U.S. mesh—so separations were inefficient by modern chromatographic standards. However, this is a problem that is easily remedied. A more serious problem from an analytical point of view is attributable to domains of "unsatisfied" ion exchange sites within the retardation resin. Ideally, the polyacrylate snake should provide an exact electrostatic balance for the sites on the anion exchange resin cage. Under this circumstance the absorption of electrolyte can be expected to be totally reversible; entering the resin to "unzip" the one polyelectrolyte from the other, washing out again with water. In practice this is found not to be the case. In any real ion retardation resin prepared by the Hatch method there are invariably domains of both cation and ion exchange sites that are not satisfied by their polyelectrolyte counterpart and must therefore be satisfied by mobile counterions. These domains amount to only a few percent of the total sites in the resin, which is a minor problem in the intended use, namely, the processing of high loadings of concentrated feeds. But in an analytical context, such an imbalanced retardation resin exchanges the typically minute amount of ions injected. As a result they are irreversibly bound and are not removed by a water eluent. If a way can be found to prepare resins wherein there is an exact balance of one polyelectrolyte with the other while still retaining selectivity, then ion retardation could be an important part of the IC repertoire of methods.

Small and co-workers[15] have used very weak base resins in conjunction with a water eluent to resolve mixtures of anions. The weak base resin, poly-vinylpyridine of colloidal dimensions, is immobilized on the surface of a sulfo-nated substrate particle by the familiar surface agglomeration technique. Anions are separated as their acids, so as a preliminary to an actual analytical run the sample is first "conditioned" by passing it through a small column of cation exchanger in the hydronium ion form.

The partitioning process is the reversible salt formation between the weakly basic sites of the resin and the acids, for example

$$\text{Resin}-\text{Py}: + \text{HA} \rightleftharpoons \text{Resin} - \text{Py:HA} \qquad (5.21)$$

where Py: denotes the pendant pyridine groups on the resin. This process is selective in that acids display different elution times (Figure 5.10). The eluted acids are detected conductometrically.

An interesting aspect of this separation is the pronounced temperature de-pendence of k-primes—the lower the temperature the higher the k-prime. This correlates with the increase in the ionic product of water that accompanies an increase in temperature and a like dependence of salt hydrolysis on temperature.

When a resin of stronger basicity is employed, water is not an effective eluent since the left-hand side of the equilibrium of (5.21) is too favored. In that case elution with a stronger "base" than water is called for and a dilute solution

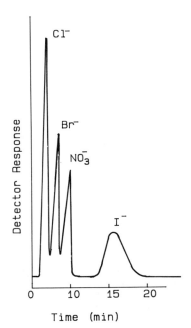

FIG. 5.10. Separation of anions as their acids on a weak base resin containing pyridine functional groups. Water eluent.[15]

of pyridine has proved to be effective. The eluted acids are detected conductometrically by the increased conductivity of the pyridinium salts above the base
line of the weakly conducting pyridine solution.

An analogous type of separation was attempted for cations. In this case the
stationary phase was prepared by agglomerating a layer of finely divided carboxylic acid resin (Dowex CCR-2) onto substrate particles of Dowex-1 in the bicarbonate form. The mobile phase was 0.01 M carbonic acid, prepared by passing
0.01 M $NaHCO_3$ through a large bed of the hydrogen form of Dowex 50WX8.
This system gave good separations of sodium and potassium, but a disturbing
problem appeared when the amounts of the analytes in the sample were reduced.
Then a negative response in the base line, before and after the sample peaks,
grew until the sample peaks actually became troughs (Figure 5.11). At first
puzzling, this effect was finally recognized as having the same origin as the
conductance dip that one sees in the initial stages of the conductometric titration
of a weak acid such as carbonic acid. This successful accounting of the negative
peaks in this cation analysis system also cleared up a mystery in suppressed
conductometric detection that had perplexed us for some time. This is fully
discussed in Chapter 7, Section 7.3.4d, while the theoretical basis of the effect is
provided in Appendix C.

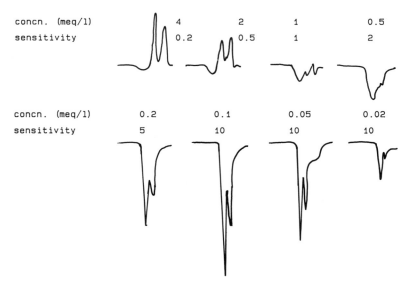

FIG. 5.11. The separation of sodium and potassium on a pellicular carboxylic acid
resin. Water eluent; analytes injected as their hydroxides; conductometric detection.
Shows the effect on peak response as the concentration of Na^+ and K^+ in the sample
was reduced from 4 to 0.02 meq/liter in each. The relative detector sensitivity is listed
below the concentration. (J. W. Pischke and H. Small, unpublished results.)

This peculiarity of carbonic eluents is clearly a drawback in that it compromises the accurate quantitation of small amounts of injected analytes. It could be avoided by employing a more weakly acidic resin as stationary phase and water as the mobile phase—analogous to polyvinyl pyridine in the anion case. However, no such resin is readily available.

Although the methods cited have their various drawbacks, that does not diminish the incentive to develop water–eluent-based IC methods. While ion exchange chromatography is a powerful means for the separation of ions, the detection of analyte is often complicated by the presence of electrolyte that must necessarily be present in the mobile phase (Chapter 7). A method that employed only water as eluent would have a number of features to recommend it. In the first place water is an excellent medium for applying conductometric detection, one of the most widely applicable detection methods. Using water as mobile phase eliminates the problem of eluent make-up. While this at first glance may seem to involve minor expense or convenience it is not a trivial matter when IC is intended for use in a continuous, round-the-clock fashion. Then the advantages of a system that uses only water as eluent become apparent. These can be brought out by considering the hypothetical system depicted in Figure 5.12. Water eluent is pumped continuously to the separator column by way of a sample injection valve and a short precolumn that conditions the sample, that is, converts it to a common anion or cation form as the situation demands. After separation the electrolytes are detected conductometrically and then removed in a small mixed

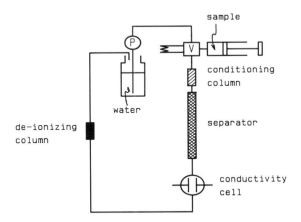

FIG. 5.12. Concept for the continuous analysis of electrolytes using water as eluent. Water is circulated around the system. Samples are injected and a conditioning bed of ion exchanger converts all salts to a common cation form for anion analysis or a common anion form for cation analysis. The salts are separated in the column and detected conductometrically. The effluent then passes through a small deionizing bed that effectively removes the analyte electrolytes and purified water is returned to the eluent reservoir.

bed deionizer. The purified water is then returned to the reservoir for reuse. Detection can be expected to be very sensitive since the background is essentially deionized water. Because the amount of sample injected is typically in the nanogram to microgram range, the preconditioning and deionizing beds should have a long lifetime before they exhaust, so it is clear how such a system should have great potential for unattended operation. One can therefore expect research into water-based eluent systems to continue with some vigor.

REFERENCES

1. R. M. Wheaton and W. C. Bauman, Non-Ionic Separations with Ion Exchange Resins, *Ann. N.Y. Acad. Sci.* **57** (Art. 3), 159–176 (1953).
2. R. M. Wheaton and W. C. Bauman, Ion Exclusion. A Unit Operation Utilizing Ion Exchange Materials, *Ind. Eng. Chem.* **45**, 228–233 (1953).
3. D. Reichenberg and W. F. Wall, The Adsorption of Uncharged Molecules by Ion-Exchange Resins, *J. Chem. Soc.* 3364–3373 (1956).
4. A. J. P. Martin, Some Theoretical Aspects of Partition Chromatography, *Biochem. Soc. Symp.* **3**, 4–20 (1950).
5. H. Small and D. N. Bremer, Sorption of Nonionic Solutes on Ion Exchange Resins Bearing Amphiphilic Counter Ions, *Ind. Eng. Chem. Fund.* **3**, 361–367 (1964).
6. R. N. Sargent and W. Rieman III, Salting-out Chromatography I. Alcohols, *J. Phys. Chem.* **61**, 354–358 (1957).
7. R. N. Sargent and W. Rieman III, Salting-out Chromatography II. Amines, *Anal. Chim. Acta.* **17**(4), 408–414 (1957).
8. R. N. Sargent and D. L. Graham, Reversible Sorbents for Organic Acids and Amines, *Ind. Eng. Chem. Process Design Dev.* **1**(1), 56–63 (1962).
9. V. T. Turkelson and M. Richards, Separation of the Citric Acid Cycle Acids by Liquid Chromatography, *Anal. Chem.* **50**, 1420–1423 (1978).
10. C. J. Pedersen, Cyclic Polyethers and Their Complexes with Metal Salts, *J. Am. Chem. Soc.* **89**, 7017–7036 (1967).
11. E. Blasius, K.-P. Janzen, W. Adrian, G. Klautke, R. Lorscheider, P.-G. Maurer, V. B. Nguyen Tien, G. Scholten, and J. Stockemer, Preparation, Characterization and Application of Complex Forming Exchangers with Crown Compounds or Cryptands as Anchor Groups, *Z. Anal. Chem.* **284**, 337–360 (1977).
12. R. M. Cassidy, The Separation and Determination of Metal Species by Modern Liquid Chromatography, in *Trace Analysis,* Vol. 1 (J. F. Lawrence, ed.), Academic Press, New York (1981).
13. K. Kimura, H. Harino, E. Hayata, and T. Shono, Liquid Chromatography of Alkali and Alkaline-Earth Metal Ions using Octadecylsilanized Silica Columns Modified in situ with Lipophilic Crown Ethers, *Anal. Chem.* **58**, 2233–2237 (1986).
14. M. J. Hatch, J. A. Dillon, and H. B. Smith, Preparation and Use of Snake-Cage Polyelectrolytes, *Ind. Eng. Chem.* **49**, 1812–1819 (1957).
15. H. Small, M. E. Soderquist, and J. W. Pischke, Weak Eluent Ion Chromatography, U.S. Patent No. 4,732,686 (1988).

Chapter 6

Detection—General

6.1. INTRODUCTION

While separation is the central issue in chromatography, detection of the separated species is of comparable importance. In fact, the very pace of development of chromatographic techniques has hinged on the availability of suitable detection devices. Scott,[1] writing in 1977, made the point that "the rapid development of gas chromatography arose solely from the availability of sensitive detecting systems" and further that the widespread use of liquid chromatography was retarded by the lack of suitable detectors. So present day standards require not only the development of fast and efficient separation techniques but also that they be coupled with detectors that can provide prompt and precise measurement of the eluites over a broad range of sensitivities. In this respect ion chromatography has very much followed the same pathway; the potential of ion exchange chromatography has been considerably enhanced by the development of suitable detectors.

Detection in IC is in some respects more problematical than in most other forms of chromatography. This is not meant to imply that the problems in other areas have been trivial nor that IC detectors are fraught with problems, but rather that the detection problems of IC are complicated by the chemistry of the separation processes to a greater extent than most other chromatographic techniques. The separation and the detection parts of IC are in fact so intimately entwined that the constraints and requirements of both must be kept prominently to the fore while developing an IC method.

Problems also arise when trying to classify IC detectors, although this has been done successfully for conventional LC. Whereas the terms "bulk property detection" and "solute property detection" delineate the two approaches to detection in LC, ambiguities arise when these terms are extended to include IC detection. For example, cases abound in IC of UV detectors being used in either a bulk property or solute property mode while, in LC practice, a UV detector is commonly cited as a prime example of a solute property detector.

Because of these difficulties, rather than force IC detectors into one category or the other, I will develop the principles that unify the diverse methods and point out from time to time the classification to which a detector may be ascribed if that seems appropriate.

6.2. DETECTION IN CHROMATOGRAPHY

6.2.1. The Elements of Detection

In elution development chromatography the eluites of interest appear at the column outlet among a preponderance of other species. It is the task of the detector to measure, with considerable precision, the concentration of the eluites in this oftentimes overwhelming background of other materials that make up the eluent.

The detection procedure can be depicted as in Figure 6.1. Element A encompasses all those steps that might be described as effluent conditioning. This might involve a variety of measures; careful thermostating if the measured property is temperature sensitive, dampening of pressure or flow irregularities if the detector is sensitive to that type of disturbance, or, as we will see in conductometric detection, altering the chemistry of the effluent before it passes to the input transducer.

The function of the input transducer, B, is to convert the property—absorbance, conductance, electrochemical activity, refractive index, etc.—into an electrical signal. The transducer incorporates a detecting cell where effluent is "probed" for the property of interest, and electronics to convert property to signal. It is important to minimize the volume and control the construction of the cell so as not to seriously degrade the sharpness of the eluting solute peaks, but since each detection method has its own peculiar requirements in this regard, this issue will be pursued later as we examine each method separately.

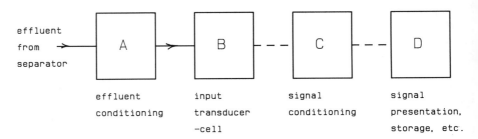

FIG. 6.1. The elements of detection in chromatography.

Element C represents various signal conditioning procedures—amplification, filtering, etc.—that are applied before the signal is passed to the final elements that will include means of recording the signal as a function of time, either visually as on a strip chart recorder or in the memory of a computer.

6.2.2. The Requirements of a Detector

A primary requirement of a detector is that it should be responsive over a wide range of concentrations of the monitored solute. Some detectors will respond to changes in solute level when its concentration is low but be prone to "saturation" effects and become relatively unresponsive when concentrations are high. The range of measured property over which the detector is responsive is termed the **dynamic range** of the detector.

In chromatography it is also desirable that the output signal be a linear function of eluite concentration and that this linearity should persist over a very wide range of concentration; this is a convenience for calibration and in the interpretation of data. The range of concentration over which this linearity applies is termed the **linear dynamic range** of the detector. There are a number of reasons why a detector may deviate from this ideal.

In the first place it may be an inherent quality of the transducer that it is not able to deliver a linear signal when stimulated by a linearly varying property. Nonlinearity can, however, be closely related to eluite chemistry. When the property of interest is not proportional to the concentration, then the input transducer, even though it may be linear in the sense just described, must necessarily deliver a nonlinear signal to the next element in the chain. For example, the conductivity of solutions of strong electrolytes is not proportional to concentration over the whole range of concentrations in which one might want to use conductometric detection. In ion chromatography, however, conductivity is proportional to concentration within the range of most interest.

If the signal from the transducer is grossly nonlinear but is a well-defined function of concentration, then element C will often be used to "operate" on the signal and generate the desired linearity. A photometer, although it is not usually thought of in this way, is an example of such a detector. The transducer, in this case a photocell of some sort, outputs a signal that is proportional to the intensity of the incident light. The intensity of the incident light, however, is dependent in an exponential way, not a linear way, on the concentration of solute interposed between the light source and the photocell (see Section 8.2.1). Photometers invariably operate on the signal and generate another property that is linear with concentration. That property is, of course, known as absorbance.

The magnitude of change of signal per unit change in concentration of analyte is often called the **response** of the detector. Since versatility is a commendable attribute in a detector, the most useful are those that are responsive to a wide variety of solutes while at the same time not displaying extreme differences in responsiveness from species to species.

The signal induced per unit amount of analyte is often referred to as the **sensitivity** of the detector. This definition is commonly used in other analytical techniques, but when used in chromatography it can lead to confusion and in some cases to grossly misleading conclusions. When used in a chromatographic context, any proper definition of the term "sensitivity" should incorporate some statement about noise. This matter will be pursued in the following sections.

6.2.3. Sensitivity, Noise, and Limits of Detectability

Since the concentrations of analytes are oftentimes extremely low, another measure of detector usefulness is its ability to discriminate between the signal from the analyte and signals from other sources in the system. The problem may be elaborated as follows. The effluent that proceeds to the detection phase comprises not only the analytes but also all the components that make up the mobile phase: solvents, electrolytes, ion pairing reagents, etc. The transducer will be responsive to all of these but of course to varying degrees. It would clearly be preferable if the transducer were responsive only to the analyte species, and in chromatography it is often possible to arrange that this be approximately so, that is, by preparing mobile phases that are effective in a separating sense but are as devoid as possible of the property to which the detector is sensitive or to which it is "tuned." When these conditions prevail, the detector is often said to be a solute property detector or selective detector. In conventional LC practice, photometers, fluorometers, and electrochemical detectors are often cited as prime examples of solute property detectors; conditions are often such that the mobile phases are essentially devoid of absorbance, fluorescence, or electrochemical activity while the analytes possess these respective properties, often abundantly. The ability of solute property detectors to detect low levels of eluting analyte is then limited by the intrinsic noise of the detector. What is meant by noise and specifically intrinsic noise?

Despite our best efforts to be precise, the measurement of any physical property or quantity is subject to a degree of uncertainty. This imprecision usually arises either from an inability to precisely control the measuring environment or from the inability of the measuring device to return exactly the same value when presented with "identical" samples for measurement. This uncertainty or variance in the system is commonly called **noise.**

Noise has been defined[2] as "any perturbation on the detector output that is not related to eluted solute." It can take various forms and have quite different origins, but whatever its origin it is a fundamental property of all real detection systems and it is the property that imposes a limit on the sensitivity of a detector or a detection system. This is graphically depicted in Figure 6.2, which illus-

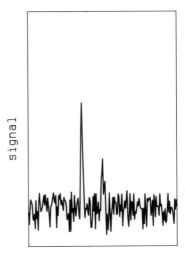

FIG. 6.2. Detection of chromatographic peaks is compromised by a noisy background signal.

volume

trates how a diminishing signal from a chromatographic peak eventually becomes practically indistinguishable from the base-line noise.

Figure 6.3 illustrates three categories of noise in relationship to a chromatographic peak. The high-frequency hash (type 1) usually has its origin in detector electronics. In a well-designed detector this noise is of low amplitude and will only become manifest at highest sensitivity settings. Furthermore, it is amenable to filtering since it usually has a very different "frequency" from most of the chromatographic peaks. The second type of noise has about the same

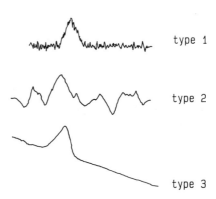

FIG. 6.3. Types of noise in detection in chromatography.

"frequency" as the analyte peak and may be caused by deficiencies in detector components. When both these types of noise originate within the detector itself the term "intrinsic noise" is appropriate.

The third type of noise, termed drift, while it may be a deficiency of the detector is much more likely to be caused by factors that are extraneous to the detector. It might be caused by slow equilibration between the stationary and mobile phases or a drift in an important external factor such as the temperature of the system.

Inadequate control of important system variables may also be the seat of the second type of noise. These extraneous sources of noise can become the dominant factor limiting the sensitivity of a chromatographic technique and are indeed the most vexatious for certain methods of ion chromatography.

The magnitude of this type of noise is intimately associated with the composition of the ambient fluid carrying the solute into the detector and most vitally dependent on the extent to which that fluid possesses the property to which the detector is sensitive. So important are these matters to the field of ion chromatography that their general features will now be examined in some detail.

6.2.4. Mobile Phase as a Source of Noise and Limitations on Detectability

The feed to a chromatographic detector contains many species in addition to the solutes of interest. Assume that one or more of these ambient species possesses the property to which the detector is sensitive. [N.B. The language "possess the property to which the detector is sensitive" is deliberately chosen rather than language such as "possesses the same property as the eluite." This choice accommodates that important group of "indirect" methods wherein the eluites lack the property examined by the detector (see Section 8.2.4).]

Recognizing that the control of the systems is imperfect, let us assume a resulting measure of uncertainty as depicted in the plots of Figure 6.4. To give an example of how this uncertainty might arise: If the property examined is temperature sensitive then the random noise could be a direct result of the random fluctuations in temperature of the fluid reaching the detector. Fluctuations in the concentrations of a species bearing the subject property would have a similar effect.

The three traces in Figure 6.4 are what one would expect from three different experiments wherein the detector receives different levels—concentration perhaps—of the property-bearing substance; or alternatively three streams where the property being measured is present to different degrees denoted as 1000, 500, and 100 arbitrary units.

Noticeable of course is the noise that accompanies what, if circumstances were ideal, would be absolutely flat unvarying signals. But even more significant is how the *breadth* of the noise "envelope" diminishes as the magnitude of the

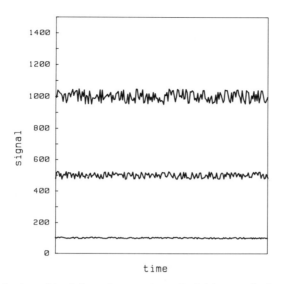

FIG. 6.4. How the breadth of the noise envelope diminishes as the background signal decreases.

signal decreases. Over a wide range of signal levels the absolute noise will be proportional to the magnitude of the signal, simply because the perturbation is a fixed percentage of the base signal. At the lowest signal levels this relationship between noise and signal will usually deviate from proportionality as the detector adds a fixed low level of intrinsic noise.

The effects represented by Figure 6.4 have a profound implication for chromatography for it shows the important benefit to be gained from working with low ambient signal levels. Figure 6.5 puts this in a chromatographic context by showing how the noisy background compromises the ability to accurately measure the analyte peak while a tenfold lowering of the background signal has an obviously beneficial effect on detectability.

This argues in favor of using solute property detectors wherever possible. It also conveniently explains the origin of the term "bulk property detector." The terminology arose because the detector examined a property possessed by the bulk of the eluent, the solvent in other words. The refractometer was an early example of a bulk property detector and its sensitivity limitations are well known. The above discussion and Figure 6.5 illustrate the basis of the often cited disadvantage of bulk property detectors, namely, their relatively poor sensitivity and limited dynamic range.

This treatment also shows how important it is to consider noise when evaluating and comparing chromatographic methods. Thus, while two detection systems might provide the same response per unit amount of analyte, as repre-

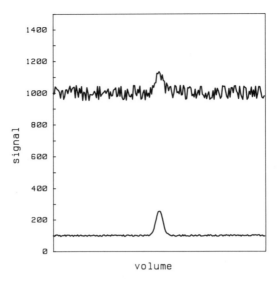

FIG. 6.5. How detection of a chromatographic peak is beneficially affected by lowering the background signal.

sented in Figure 6.6, their abilities to *detect* small amounts of analyte are clearly very different. It is only reasonable that the term "sensitivity" should convey, in the truest possible way, this ability to discriminate between signal and noise. Snyder and Kirkland[3] state: "It is important when comparing detectors that the sensitivities at equal noise levels be used." while Scott[4] defines "detector

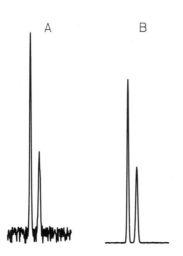

FIG. 6.6. In chromatography, response can be a misleading measure of sensitivity. In chromatogram A the species display a larger response than in chromatogram B, but the level of detectability, which is the true measure of sensitivity, will be lower in case B because of the much lower background noise. (It is assumed that the amounts of analyte in A and B are the same.)

sensitivity" as "the minimum mass/unit time or minimum concentration of solute in mass/unit volume passing through the detector that can be discerned from the noise."

In ion chromatography a number of detectors have been used in both the solute property and bulk property modes. It would seem clear, therefore, when comparing the sensitivity in such cases, that figures on noise must be included for the comparisons to have any validity. Yet this is frequently not done. The violations of this requirement are particularly prevalent in conductometric detection in IC where *response* is often given as the sole measure of a method's sensitivity. In some cases the response per unit amount of analyte is actually greater in the less sensitive method (Chapter 7, Section 7.5). In this instance, treating sensitivity and response as being synonymous is incorrect.

There are circumstances when comparison of response is an adequate and legitimate means of comparing sensitivities. Comparing responses as a means of comparing sensitivities for various analytes *within a single method* would be one example of such a use. Here it is assumed that the comparison is made *at equal noise level.*

6.2.5. Definition of Detectability

Since the relative magnitudes of the signal and the noise determine sensitivity, it is logical to use the signal-to-noise ratio to evaluate the limit of detectability of the system. In turn it is the convention to express the limit of detectability as that level (concentration or mass) of solute that produces a signal that in average amplitude is twice the average amplitude of the noise (Figure 6.7). This limit is known by various names; minimum detectable limit (MDL) and limit of detection (LOD) are two in common use.

6.2.6. Detectability in Chromatography

In chromatography "minimum detectable limit" is often ill defined and loosely used. This is in great part due to the fact that chromatographic performance as well as the detection system determines the ultimate sensitivity of the

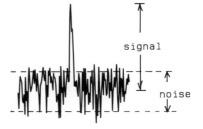

FIG. 6.7. Detectability. Minimum detectable limit (limit of detection) is often defined as that level of analyte that produces a signal that in average amplitude is twice the average amplitude of the noise.

system. To use one simple example: A species under one set of chromatographic conditions that gives a sharp peak would be detectable at lower levels than at conditions yielding a broader peak.

Claims for levels of detectability in chromatography should be examined with considerable caution, particularly if the chromatographic conditions and operating parameters are ill defined.

REFERENCES

1. R. W. P. Scott, *Liquid Chromatography Detectors,* p. 1, Elsevier, Amsterdam (1977).
2. Reference 1, p. 15.
3. L. R. Snyder and J. J. Kirkland, *Introduction to Modern Liquid Chromatography,* 2nd. edition, p. 129, Wiley-Interscience, New York (1979).
4. Reference 1, p. 18.

Chapter 7

Conductometric Detection

7.1. INTRODUCTION

Conductometric detection, at one time relatively neglected, is now one of the major detection methods in the chromatographic analysis of ionic species. As a means of detection, conductance monitoring has a number of features to recommend it:

1. Electrical conductivity is a universal property of ionic solutions.
2. At the low concentrations typical of chromatography, conductivity is proportional to concentration.
3. The detector, by virtue of its simplicity, is robust and capable of prolonged trouble-free operation.
4. Since the cell is amenable to virtually limitless miniaturization, any inefficiency attributable to detector cell volume can be easily eliminated.

Despite these obvious merits, conductance monitoring displayed a serious deficiency in early attempts to apply it to ion exchange chromatography, the most powerful method for the separation of ionic species. Since the mobile phases of ion exchange chromatography are necessarily ionic and therefore often highly conducting, the conductivity detector applied in this context was a bulk property detector. It therefore suffered from the major limitation of that type of detector in being relatively insensitive to the presence of small amounts of eluite ions in the high ambient levels of eluent species. Consequently, in the early application of ion exchange chromatography and later of HPLC to ionic species, a property other than electrical conductance was monitored wherever possible.

If the eluite ions had useful optical properties, that is, absorbance in the UV or visible, then photometers were an ideal choice and if this were not the case it was occasionally possible to arrange some suitable postseparation chemistry to generate chromophores. The classic work of Stein and Moore on the ion exchange separation of amino acids coupled with postseparation addition of ninhydrin to develop measurable color is an important early example of such an approach and one that is still widely used.

While many organic ions were strong UV absorbers and therefore amenable to very sensitive detection by UV photometers there were a great many that were not. In addition there were many common inorganic species both anionic and cationic that were nonchromophoric and consequently thought not to be amenable to the chromatographic approaches that had been developed for organic materials. By the early 1970s, the application of chromatography to inorganic analysis was relatively insignificant when compared to the chromatography of organic species that had benefited so greatly from the advances in gas and later in high performance liquid chromatography. It was yet another example of progress impeded through lack of a suitable detector since the separation potential of ion exchange was already well established.

In 1971, Small, Stevens, and Bauman, working at the Dow Chemical Company, invented a means of coping with the high background conductance of the eluents of ion exchange chromatography and in effect transformed a conductivity cell from a bulk property detector to a solute property detector. They accomplished this by adding a second column of ion exchanger between the separator and the conductivity cell that either removed the eluent ions or at least greatly lowered their conductance while at the same time it enhanced the conductance of the eluites in most cases. The procedure became known as eluent suppression and was first reported in 1975.[1-3] The Dow Chemical Company was granted several patents covering eluent suppression and related methods and in turn granted to the Dionex Corporation a license to manufacture and market instrumentation using this patented technology.

In 1979 two groups of workers reported a method of inorganic anion analysis that used an ion exchange column coupled directly to a conductivity detector.[4,5] By attention to various properties of the chromatographic system, notably the capacity of the separating columns, they had improved the sensitivity of a conductometric detector when used in the bulk property mode. Several other instrument companies, barred from the suppression technology by the Dow and Dionex patents, adopted the nonsuppressed approach and have marketed instrumentation using it. This method became known under a variety of names: "single column IC" (SCIC), which has been used to distinguish it from the suppressed technique that, in its earliest form, used two columns; "direct conductometric detection," since the column effluent was passed directly to the conductivity cell; and "nonsuppressed IC," the term used in this book.

These two approaches to conductometric detection will be discussed in detail later in the chapter, but first a discussion of some of the fundamentals of electrolyte conductance is in order.

7.2. ELECTROLYTE CONDUCTANCE—THEORY AND MEASUREMENT

7.2.1. Strong Electrolytes

When electrolytes dissolve in solvents of high dielectric constant such as water they dissociate into their constituent ions and the solutions are electrically conducting. The electrical properties of the solution obey Ohm's law in that the resistance to current flow may be defined as

$$V = iR \tag{7.1}$$

where V is the voltage and i is the current that flows through an element of solution. The resistance R is a function of temperature and the concentration of the electrolyte.

The resistance of an element of solution is proportional to its thickness, l, and inversely to its cross-sectional area, A.

A **specific resistance, ρ,** may therefore be defined as

$$R = l\rho/A \tag{7.2}$$

The several ions of a solution conduct, so they can be thought of as conductors in parallel, with the net resistance R of the composite being related to the separate resistances R_1, R_2, R_3:

$$1/R = 1/R_1 + 1/R_2 + 1/R_3 \tag{7.3}$$

Since the reciprocal of resistances are additive, electrolytes can be conveniently treated by using the conductances or conductivity, which is the reciprocal of resistivity.* The **specific conductance,** denoted κ, is therefore

$$\kappa = l/\rho = l/AR \tag{7.4}$$

The conductance of the ions produced by one gram equivalent of electrolyte at any concentration can be evaluated by considering a cell with electrodes placed one centimeter apart but of sufficient area as to just enclose the whole volume containing the one equivalent of electrolyte. The conductance of such an assembly is termed the **equivalent conductance** and is denoted Λ. Equivalent conductance Λ and specific conductance κ are related in the following way.

The area (cm²) of a cell of 1 cm thickness that will enclose one equivalent of electrolyte whose concentration is C equivalents per liter is simply $1000/C$, where C is the concentration of the electrolyte in equivalents per liter. Since the electrodes of this hypothetical cell are, by definition, 1 cm apart, it follows that

---

*The unit of electrical conductance is the mho or siemens and is the conductance (reciprocal resistance) of a conductor in which a potential difference of 1 V maintains a current of 1 A.

the conductance per square centimeter is simply the specific conductance κ. The total conductance of the cell of area A is therefore Aκ. By definition this conductance is the equivalent conductance Λ. Substituting for A gives

$$\Lambda = 1000 \; \kappa/C \qquad (7.5)$$

This is a very important expression for conductometric detection.

In the case of weak electrolytes the concentration of contributing ions is of course quite different from the formal concentration.

Conductance would be an exact linear function of concentration if Λ was an intrinsic property of an electrolyte and therefore not dependent on its concentration. It has been known for some time that for strong electrolytes Λ depends on concentration in a manner that may be described by the expression

$$\Lambda = \Lambda_0 - B\sqrt{C} \qquad (7.6)$$

where B is a constant.

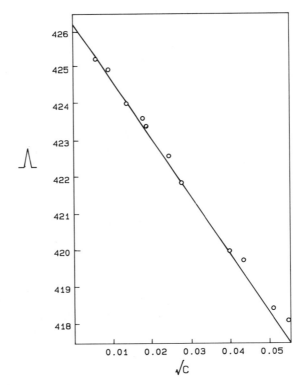

FIG. 7.1. Equivalent conductivity of hydrochloric acid in very dilute aqueous solution at 25°C. The points represent experimental data; the solid line is calculated from the Onsager limiting law. (From Ref. 6 with permission.)

This predicts that at sufficiently low concentration the equivalent conductance approaches a limiting value that is independent of the degree of dilution of the system. At this point the constituent ions of the system are sufficiently far apart that they exert no effect on one another and act as independent noninterfering conductors. When Λ becomes independent of concentration it also follows [equation (7.5)] that the specific conductance of the electrolyte solution is a linear function of concentration. The experimental and a theoretically derived dependence of equivalent conductance and concentration for a typical strong electrolyte are illustrated in Figure 7.1. It is evident from Figure 7.1 that for solution concentrations below 10^{-3} M an insignificant error is introduced by assuming conductances of electrolyte and their constituent ions to be equal to their **limiting equivalent conductance,** that is, their equivalent conductance at infinite dilution. Theoretical calculations in IC invariably use values for limiting equivalent conductance, a practice that is justified since most applications of conductometric detection in IC involve very low concentrations of eluent or eluite.

The additivity of ion conductances is sometimes referred to as the Kolrausch law of independent ion mobilities and is a very useful approximation in theoretical calculations of the conductance of solutions. Thus the conductance of any dilute electrolyte can be calculated simply by summing the contributions from its constituent ions:

$$\Lambda_{NaCl} = \lambda_0^{Na} + \lambda_0^{Cl} \tag{7.7}$$

Values for limiting equivalent conductances are available from many sources. A tabulation of values is provided in Appendix B.

7.2.2. Weak Electrolytes

When a weak electrolyte such as acetic acid (HA) is dissolved in water it dissociates partially into its constituent ions:

$$HA \rightleftharpoons H^+ + A^- \tag{7.8}$$

The equilibrium constant K_a defines the dissociation:

$$K_a = [H^+][A^-]/[HA] \tag{7.9}$$

If the formal concentration of acid is denoted C, then the concentrations of the various species are related as follows:

$$[H^+] = [A^-] \tag{7.10}$$

$$[HA] = C - [H^+] \tag{7.11}$$

Substitutions in (7.9), (7.10), and (7.11) lead to the expression

$$[H^+]^2 + K_a[H^+] - K_a C = 0 \tag{7.12}$$

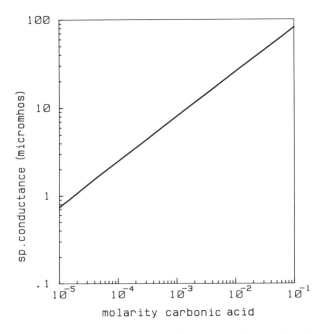

FIG. 7.2. The calculated specific conductance (μmho/cm) of carbonic acid as a function of concentration.

Solving the quadratic for a positive root of $[H^+]$ gives

$$[H^+] = [-K_a + (K_a^2 + 4K_aC)^{1/2}]/2 \qquad (7.13)$$

Since $[A^-] = [H^+]$, equation (7.5), combined with the limiting equivalent conductances of the anion and hydronium ions, permits calculation of the conductance of the weak acid at any concentration. The results of such calculations for carbonic acid, an important case for IC, are shown in Figure 7.2.

7.2.3. Measurement of Conductance

While the concept of electrical resistance or conductivity of electrolytes is a simple one, the accurate measurement of these quantities is not trivial. Polarization effects and double layer charging effects at the electrodes of the conductivity cell are complications of the measuring process that are alleviated by applying alternating (typically 1000 Hz) or pulse potentials.[7,8] Measurements of the current through the cell are made when charging currents have decayed to very low levels. The user of IC is normally not confronted with the problems of conductance measurement since that is more in the domain of the electrical engineer who designs the cell and accompanying electronics. Weiss[8] has provided a useful summary of some of the considerations in linking conductometric measurement and microprocessors.

7.3. CONDUCTOMETRIC DETECTION: SUPPRESSED

7.3.1. Eluent Suppression

The principle of eluent suppression[1] may be conveyed by a simple example, the separation and detection of sodium and potassium ions. We have seen how separation of these two ions may be effectively accomplished on a column of strong acid cation exchanger using a mineral acid solution as mobile phase. By applying conductometric detection directly to the effluent from the column it is possible to detect the sodium and potassium eluites since they displace equivalent amounts of the more conducting hydronium ions from the eluent. This results in a drop in conductance within the eluite bands, although it is common practice to reverse the polarity of the output signal so that the chromatograms are displayed in the conventional way as upscale deflections. At any rate a change in conductance results.

If the concentration of eluite is low then the compositional change within the eluite band can be quite small relative to the ambient concentration of hydronium ions. The situation represented by Figure 7.3 can then arise where the noise

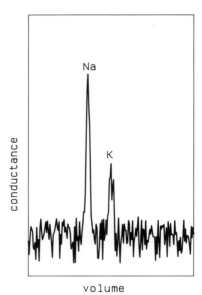

FIG. 7.3. Noise interfering with the accurate conductometric measurement of peak heights (computer simulation).

in the signal from the background severely compromises the accurate measurement of the eluite peaks. This is a good example of the major drawback of a bulk property detector.

Since reduction of noise is the key to greater sensitivity and noise is proportional to background signal (Chapter 6), then it follows that elimination of the background electrolyte should yield a concomitant reduction in the level of noise. By adding a second column containing a strong base resin in the hydroxide form the acid eluent is effectively removed

$$\text{Resin-OH}^- + \text{HCl} \rightarrow \text{Resin-Cl}^- + \text{H}_2\text{O} \qquad (7.14)$$

while the eluite species appear as bands of sodium and potassium hydroxides in a background of essentially pure water of much lower conductance than the eluent (Figure 7.4):

$$\text{Resin-OH}^- + \text{NaCl} \rightarrow \text{Resin-Cl}^- + \text{NaOH} \qquad (7.15)$$

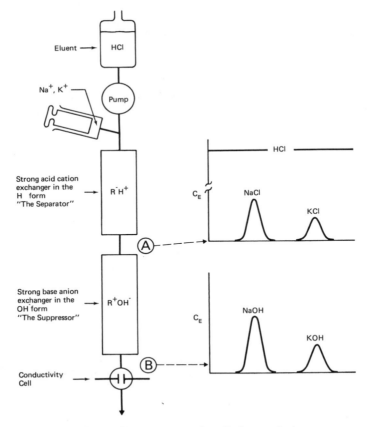

FIG. 7.4. Eluent suppression. Cation analysis.

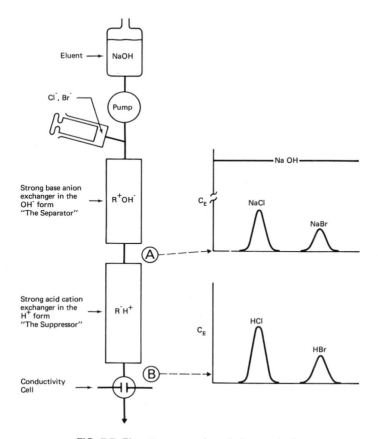

FIG. 7.5. Eluent suppression. Anion analysis.

An analogous scheme for the separation and detection of anion mixtures is depicted in Figure 7.5. In this example sodium hydroxide is the eluent and a strong base resin in the hydroxide form is the stationary phase. The eluites elute in a background of highly conducting sodium hydroxide, which is removed by directing the effluent through a second column filled with a strong acid resin in the hydrogen form. In a similar way the conducting eluent is replaced with essentially deionized water with the concomitant reduction in noise. The eluite species appear as bands of their corresponding acids, which, if they are strong enough, are highly ionized and conducting. It is also worth noting that the presence of the highly conducting hydronium ion in the analyte band enhances the analyte signal still further relative to the base-line noise.

Eluent suppression is in essence a means of improving signal-to-noise ratio; it does so by removing or suppressing a major source of noise, the signal contributed by the mobile phase conductance. Solving one problem can often lead to another, in this instance suppressor exhaustion.

7.3.2. Exhaustion and Regeneration of Column Suppressors

The suppressor column differs in an important way from the separator. While the latter, in principle, continues to provide the same separation power regardless of the amount of eluent passed through it, the suppressor's ability to neutralize eluent is limited and it eventually becomes exhausted and must be regenerated or replaced. This need to regenerate or replace a column as part of the routine made early versions of IC unique compared to most chromatographic methods. In present day IC, column suppressors have been largely eliminated by the use of continuously regenerated membrane suppressors, but since many of the factors that are important to the operation of column devices pertain also to these more recent techniques, a detailed treatment of column-based suppressors is appropriate.

Clearly a too frequent need to regenerate the suppressor would be a drawback, so the practicality of suppression depends in part on devising means of making this regeneration step as unobtrusive as possible.

The problem of suppressor exhaustion may be articulated in the following example. Consider the elution through an anion exchanger separating bed of a series of monovalent anions A_1, A_2, A_3, ... A_n, where A_1 is the least and A_n the most tightly held species (Figure 7.6). The separating column is assumed to

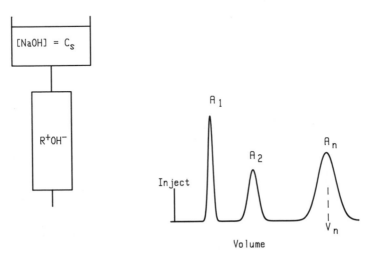

FIG. 7.6. Elution of a set of anions A_1–A_n through an anion exchanger using sodium hydroxide at concentration C_s as eluent.

have a volume of 1 ml. The volume of eluent required to elute a single sample is the volume (V_n) required to elute the most tightly held anion and is given by the expression

$$V_n = K_{OH}^{An} (C_r/C_s) \qquad (7.16)$$

where K_{OH}^{An} is the selectivity coefficient expressing the separating column's preference for the anion A_n^- relative to OH^- the displacing ion, C_r is the specific capacity of the separator resin in milliequivalents per milliliter, and C_s is the concentration of sodium hydroxide in the same units [see equations (2.8) and (4.7)]. More correctly, the volume required to elute A_n completely is greater than V_n since V_n is the volume from injection to peak maximum. Complete elution implies eluting until the response for a species has returned to the base line. Furthermore, equation (7.16) neglects to include the void volume of the bed, which is assumed to be small compared to V_n. However, any conclusions drawn from what follows will not be invalidated by the approximations implicit in (7.16).

Rearranging this equation gives

$$V_n C_s = K_{OH}^{An} C_r \qquad (7.17)$$

so that the term on the left, $V_n C_s$, represents the number of milliequivalents of eluent required to elute a single sample. This in turn is the *minimum* amount of suppressor capacity required to neutralize or suppress this amount of eluent. Hence, the volume of suppressor required for the elution of N samples through 1 ml of separator bed is given by the expression

$$V_{sup} = NK_{OH}^{An} (C_r/C_{sup}) \qquad (7.18)$$

where C_{sup} is the specific ion-exchange capacity of the suppressor in meq/ml.

How large can V_{sup} become? The number of samples, N, must necessarily be large for the method to be useful. From the data of Table 4.2 it is evident that for many ions K_{OH}^{An} can be very large indeed since the hydroxide ion has, relatively speaking, a low affinity for the resin. It is clear, therefore, that using conventional high-capacity ion exchangers in the separator and suppressor would require that the volume of the suppressor be very large since with conventional resins C_r and C_{sup} would be approximately equal. A suppressor of large volume is undesirable for at least two reasons. In the first place the efficiency of the separation is severely degraded through band broadening in the very large void space of a massive suppressor column. Secondly, the need to displace the void volume of this column introduces large dead time intervals to the analysis. Consequently, an early objective in the development of suppressed IC was to devise means of making the volume of the suppressor bed as small as possible, preferably equal to or less than the volume of the separator column but not greater than ten times, as an arbitrarily chosen upper limit.

Equation (7.18), as well as expressing the suppressor problem, also suggests solutions. They are as follows:

1. Make the ratio C_r/C_{sup} as small as possible. This is accomplished by using resins of high specific capacity in the suppressor column and resins of very low specific capacity in the separator column. When suppressed IC was in its initial research stages the first requirement was easily met by the ready availability of commercial resins of high capacity. It was the need for low-capacity separator resins that led to the invention and development of low-capacity varieties of resins such as the "surface sulfonated" cation exchangers[9] and "surface agglomerated" anion exchangers.[1,10,11]

2. Make the selectivity coefficient small. Sodium hydroxide from one point of view is the ideal eluent in that its reaction with the suppressor produces water, the ideal effluent. On the other hand, the hydroxide ion lacks potency as a displacing ion (Table 4.2) so that high concentrations of hydroxide are required to displace high-affinity anions which, in turn, imposes the penalty of early suppressor exhaustion.

Today, suppressed IC technology makes use of membrane devices that suppress continuously without need of interruptions. Nevertheless, the performance of these devices is so closely dependent on the nature of the eluent and on the concentration and rate at which it is presented to them, that the following guidelines for the choice of eluent are as pertinent to membrane suppressors as they are to the column devices for which they were initially developed.

7.3.3. Eluents for Suppressed IC

It is evident from (7.18) that the volume of suppressor necessary to accommodate a certain number of sample injections may be made smaller by employing displacing ions with a high affinity for the separating resin relative to the sample ions. Or stated in another way, the higher the affinity of the displacing ion for the resin, the lower is the concentration necessary to elute a sample, thus reducing the load on the suppressor and prolonging its lifetime. At first glance this suggests a large choice in eluent ions, but the choice is narrowed considerably by the following requirements that the eluent must also meet:

1. The eluent conductance must be amenable to suppression by some convenient reaction of the eluent ions in the suppressor device.
2. The sample ions must not be subject to an adverse reaction in the suppressor. Examples of adverse reaction would be precipitation or conversion to a weakly dissociated product.
3. The affinity of the displacing ion must not be so high that the sample ions elute too rapidly without adequate resolution.

TABLE 7.1. Eluents for Anion and Cation Analysis by Suppressed Conductometric Detection[a]

Eluent	Eluent ion	Suppressor resin	Products of suppressor reaction
Anion analysis			
NaOH	OH^-	R^-H^+	$R^-Na^+ + H_2O$
sodium phenate	PhO^-	R^-H^+	$R^-Na^+ + H_2O$
$Na_2CO_3/NaHCO_3$	CO_3^{2-}/HCO_3^-	R^-H^+	$R^-Na^+ + H_2CO_3$
Na_2 glutamate	$Glut^{2-}$	R^-H^+	$R^-Na^+ + R^-Glut \cdot H^+$
Cation analysis			
HCl	H^+	R^+OH^-	$R^+Cl^- + H_2O$
$AgNO_3$	Ag^+	R^+Cl^-	$R^+NO_3^- + AgCl$
$Cu(NO_3)_2$	Cu^{2+}	RNH_2	$RNH_2 \cdot Cu(NO_3)_2$
pyridine·HCl	PyH^+	R^+OH^-	$R^+Cl^- + Py + H_2O$
p-PDA·2HCl[b]	$p\text{-}PDAH_2^{2+}$	R^+OH^-	$R^+Cl^- + p\text{-}PDA + H_2O$

[a]Reference 11.
[b]p-PDA is para-phenylenediamine.

Even with these constraints, a number of eluents for both anions and cations have been developed for suppressed IC. Some are listed in Table 7.1 along with the appropriate suppressor reaction.

7.3.3a. Anion Analysis

From one standpoint the hydroxide ion is the ideal displacing ion—the reaction with the suppressor produces water in the mobile phase. However, because of its low position on the affinity scale, relatively high concentrations of sodium hydroxide are necessary to drive tightly bound ions, even from low-capacity ion exchangers. Consequently in the early embodiments of suppressed IC that used column suppressors, hydroxide eluents were useful only for the elution of less tightly bound monovalent ions such as F^- and Cl^- and mono-basic organic acids such as acetate and formate. For more tightly bound eluites the concentrations of hydroxide required for reasonably rapid separations caused short lifetimes in the suppressor and severely limited the applicability of hydroxide-based eluents. Today, with the development of membrane-based suppressors that can continuously neutralize quite high concentrations of eluent, hydroxide is finding favor again as an eluting ion.[12] However, the eluents developed for column suppressors are still widely employed even with these newer suppressor devices.

For more tightly bound eluite ions, especially polyvalent ions such as sul-

TABLE 7.2. Retentions of Anions on a Surface Agglomerated Resin in Sodium Phenate Eluents[a]

Anion	Elution volume (ml)		
	0.005F PhO$^-$	0.01F PhO$^-$	0.015F PhO$^-$
F$^-$	1.92	1.8	1.68
Cl$^-$	3.48	2.5	2.2
Br$^-$	7.8	4.62	3.46
I$^-$	—	—	16.3
NO$_3^-$	4.02	2.9	2.6
NO$_2^-$	4.02	2.9	2.6
IO$_3^-$	1.88	1.62	1.68
BrO$_3^-$	—	—	1.94
SO$_4^{2-}$	>34	11.6	5.76
SO$_3^{2-}$	21	—	3.88
CO$_3^{2-}$	13.5	5	3.78
CrO$_4^{2-}$	—	>25	21.8
PO$_4^{3-}$	—	—	4.72
Formate	2.3	2.06	—
Acetate	2.34	2.22	1.84
Propionate	2.64	—	—
Chloroacetate	2.74	2.22	—
Dichloroacetate	5.34	3.58	—
Trichloroacetate	—	—	10.1
Glycolate	—	—	1.84
Oxalate	—	11.6	5.76
Maleate	—	7.46	4.2
Fumarate	—	10.4	5.56
Succinate	—	—	3.66
Malonate	—	—	3.78
Itaconate	—	7.56	—
Benzoate	10.9	7.18	6.3
Ascorbate	—	—	2.1/5.88
Citrate	—	—	42

[a]Reference 1. Copyright 1975 American Chemical Society.

fite, sulfate, and phosphate, the phenate ion was an early and suitable choice as an eluent; it had a high relative affinity (Table 4.2) and it suppressed to phenol, which is a very weak and therefore poorly conducting acid. Table 7.2 summarizes the elution behavior of a number of anions in eluents of various phenate concentrations.

Although phenate eluents are superior to hydroxide eluents, unidentified impurities in the phenol reagents can build up on the separator column and steadily lower its capacity for separation. This can be alleviated by placing a small sacrificial bed ahead of the sample injection valve or by using phenol derivatives such as cyano-phenol that are less subject to the degradation that introduces impurities. However, the use of phenol eluents diminished with the development of the carbonate eluent system.[11]

The carbonate system of eluents, which employs various combinations of HCO_3^-, CO_3^{2-}, and OH^-, exploits the superior affinity of the divalent carbonate ion and the weakness of carbonic acid, the suppressed product. The superior eluting power of the carbonate ion is illustrated in Figure 7.7, which shows how 0.05 N sodium hydroxide is inferior in eluting power to a much more dilute solution (0.01 N) "spiked" with 0.002 M sodium carbonate. The higher eluting power is not its only advantage; varying the composition of the bicarbonate–carbonate–hydroxide system affords a means of varying the pH in the separator over a wider range than is possible with other eluents such as hydroxide, phenate, and borate. This property may be used to alter the elution position of certain ions whose charge is pH sensitive. All these advantages have made the carbonate eluent system the workhorse eluent in suppressed IC of anions.[13]

Amphoteric species can also be used as eluents. Small and Solc[11] employed solutions of sodium glutamate in conjunction with a hydrogen form

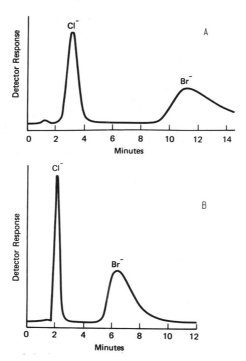

FIG. 7.7. The effect of trace carbonate on the elution power of sodium hydroxide: (A) eluent, 0.05 N NaOH; (B) eluent, 0.01 N NaOH, 0.002 N Na$_2$CO$_3$.

suppressor column to determine anions. Eluents of this type have the dual advantages of high elution power because of their higher charge and of complete removal by a cation exchange resin in the hydrogen form.

7.3.3b. Cation Analysis

The most commonly used eluents in cation IC are dilute solutions of mineral acids such as HCl, HNO_3, or H_2SO_4. Solutions in the range of $0.001-0.01$ M readily elute alkali metals, ammonium, and many organic amines from low-capacity cation exchange resins such as the surface sulfonated polystyrene type resins. The hydronium ion is an excellent choice from a suppression standpoint since it produces water in a suppression reaction with a strong base resin in the hydroxide form. However, like the hydroxide ion in anion IC it has a low ability to displace polyvalent ions such as the alkaline earth ions, thus requiring high concentrations of eluent and overburdening the suppressor. So, parallel to the practice in anion IC, eluent cations of greater affinity are used. In the early stages of suppressed IC, silver and cupric ions were employed; the silver nitrate eluent is suppressed by a combination of ion exchange and precipitation while the copper salt was removed by complexing it on a polyamine resin (Table 7.1).

In the case of silver nitrate and copper nitrate eluents, suppression is effected by stripping the electrolyte completely from the eluent. Mineral acid salts of certain organic amines, such as pyridine, aniline, and meta- and para-phenylenediamine, are effective eluents due to the high affinity of the protonated amines for the sulfonated styrene based resins—the diamines particularly so because of their dual charge in the diprotonated form. In these cases, suppression of the eluent conductance takes place not by removing the eluent but by converting it to a less dissociated and therefore less conducting form in the suppressor column, analogous to phenate and carbonate suppression in anion analysis.

For a number of years, solutions of meta-phenylenediamine in hydrochloric acid or nitric acids were the eluents of choice for suppressed IC of polyvalent metal ions such as the common alkaline earth ions. More recently, acidified diamino propionic acid (DAP) has become widely used in an analogous way to glutamate in anion analysis.

7.3.4. Suppressor Column Effects

7.3.4a. Suppression of Eluite Response

Besides its main purpose, one of the most significant effects of the suppressor is the conversion of anions to their respective acids (anion analysis) and cations to their respective bases (cation analysis), and the sensitivity of detection is related very directly to the acid and base strength of these products. Where the species give highly dissociated products, the suppressor can actually lead to an enhancement in detectability as is the case for anions of strong acids. However, for anions that form weak acids, limits of detectability rise with increasing pK of the eluite acid simply because of decreased ionization of the acid product. There is a point where suppressed IC is not an effective means for determining such species; as a rule, suppressed IC is said to work best for anions whose corresponding acids have pK values less than 7.[13] For cationic species such as amines, a similar rule applies, while for some metal ion species precipitation in the suppressor is limiting. Of course it is this selfsame relative insensitivity to detection that makes species such as carbonate and borate such excellent eluents for suppressed IC.

Bouyoucous[14] sought to overcome this drawback of suppressed IC by adding a third column of resin that converted the eluites back to a more ionized form but this approach has not been adopted routinely.

7.3.4b. Drifting Peaks

The main purpose of the suppressor column in IC is not a chromatographic one; rather, it is a reactor bed that conditions the effluent from the separator before passing it to the conductivity cell. In a minor way, the suppressor does have an influence in a chromatographic sense in that it causes some broadening of the eluite bands with a resulting loss in efficiency, but this is minimized by using small-volume, highly efficient suppressor beds. Under some circumstances, however, a suppressor column can exert a strong influence on elution behavior and can in fact change the elution order of ionic species. Furthermore, the effect can change from one chromatographic run to the next, giving rise to the phenomenon of drifting peaks. This effect was first observed by Small *et al.* in the early work on IC.[1] They noticed that under some circumstances fluoride ion would elute after chloride when selectivity considerations would dictate the reverse. Similarly, formate, acetate and chloroacetate were also observed to elute later than was expected. Moreover, the elution volume of certain eluites, acetate for instance, would decline during the lifetime of a suppressor but be restored to its original value when a freshly regenerated suppressor was put on stream. The reason for these effects becomes apparent when we examine the absorption properties of the suppressor, especially with respect to partially ionized species such as acetic or hydrofluoric acids.

Consider Figure 7.8, which illustrates a suppressor bed denoted R^-H^+, in the process of being exhausted by a basic anion eluent such as sodium hydroxide. As the eluite passes through the suppressor bed its chemistry changes in a significant way. An eluite species entering a partially exhausted suppressor enters a high pH environment where the co-ions are sodium but farther into this bed crosses the R^-Na^+/R^-H^+ boundary and enters an environment where the co-ions are hydronium ions. If X^- is the anion of a strong acid, chloride for

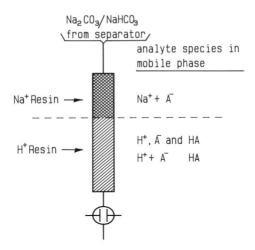

FIG. 7.8. The analyte species (A^- and HA) in the sodium and hydrogen form bands of a partially exhausted suppressor column used for anion analysis.

example, this change of environment has an insignificant effect on the rate of transport of the eluite species through this column. Donnan exclusion ensures that either sodium chloride or hydrochloric acid is prevented from entering the suppressor resin so the contribution to the total elution volume of chloride contributed by the suppressor bed is simply the void volume of that part of the system.

The retardation of species that are derived from weak acids is quite different. That part of the total elution volume contributed by the suppressor column and denoted V is given by the expression

$$V = V_v + K_1 f_{Na} V_B + K_2(1 - f_{Na})V_B \qquad (7.19)$$

The first term is simply the void volume of the suppressor. The second term describes the retention of species X^- by the sodium form of the suppressor where K_1 is the distribution coefficient of X^- between this form of the resin and the mobile phase, f_{Na} is the fraction of the bed that has been exhausted, and V_B is the total volume of the suppressor. Owing to Donnan exclusion (Chapter 5, Section 5.2.2) K_1 is very small, so this second term is negligible. The third term describes the retention of X^- and related species by the unexhausted hydrogen form of the resin. Since the X^- ion is converted to its corresponding acid at the sodium–hydrogen boundary, the elution behavior of X species through the R^-H^+ zone will depend on the strength of this acid. If the acid is strong then the predominant species will be X^-, which will be excluded from R^-H^+; that is, K_2 will be very small, and the third term will also be negligible. Thus for the anions of strong acids, as we have already seen, the only term of importance is the first and the retardation contributed by the suppressor is simply V_v.

On the other hand, if the eluite acid is weak, a fraction of the eluite will exist as the undissociated species, which, being uncharged, is unaffected by the Donnan potential and, if size is no impediment, can readily enter the hydrogen form resin phase. In this case K_2 is no longer small and the third term now becomes significant and can be the dominant factor in controlling the elution order of certain ions through the complete train of separator and suppressor.

Not only can the third term be of significant magnitude but it will also change in magnitude as the suppressor exhausts, that is, as f_{Na} changes from zero (fresh suppressor) to unity (exhausted suppressor). Thus the elution volume of species that form weak acids will be observed to drift to lower values as the suppressor becomes exhausted. A parallel case exists for cations that form weak bases, for example the ammonium ion.

In summary, suppressor effects of this type will become more pronounced

1. The higher the pK of the acid or base formed by the eluite ions
2. The larger the suppressor column relative to the separator

Peak drift is clearly a nuisance and was one of the major reasons that stimulated the development of the membrane suppressor.

The problem of drift can obviously be alleviated by using suppressors of low volume, but this incurs the penalty of more frequent regeneration. However, with modern hardware and automation techniques, frequent regeneration of very small columns might not be as bothersome a feature of suppression as it was first thought to be and column suppressors might be favored under certain circumstances.

7.3.4c. The "Carbonate Dip"

When carbonate eluents are used in suppressed IC it is common to find a negative disturbance on the base line in addition to the positive eluite signals. This is a problem if it occurs at the same position as an eluite peak, but, at the same time, one that can often be remedied by a slight alteration in elution conditions—a change in eluent strength, for example. When column suppressors are used, this disturbance has a much more bothersome feature in that, like some eluite peaks, it drifts as the suppressor is consumed. Thus, though it may not be a problem at one stage in the lifetime of the suppressor it may become so at another. The nature and the origin of the carbonate dip, as this disturbance is called, is also seated in the sorption and Donnan exclusion properties of the suppressor resin.

When the carbonate-based eluent enters the hydrogen form resin, carbonic acid is formed. Carbonic acid, being a very weak acid, is highly associated and the neutral carbonic acid species can freely enter the resin phase. Consequently, carbonic acid will not appear in the effluent to the conductivity cell for some time after a freshly regenerated suppressor is put on stream but will be retarded until the resin has absorbed its equilibrium level of carbonic acid. The volume at which breakthrough of carbonic acid takes place is given by

$$V_E = V_v + K_D V_B \qquad (7.20)$$

where V_B and V_v are the volume and the void volume, respectively, of the suppressor bed and K_D is the distribution of carbonic acid between water and the resin phase. When carbonic acid does eventually break through it will give rise to a conductance that is determined by (1) its concentration, that is, the concentration of carbonate species in the eluent; (2) the degree of dissociation of carbonic acid; and (3) the conductances of the ionic species present—predominantly bicarbonate and hydronium.

If an injection is made that gives rise to a higher carbonate level than the ambient level of carbonate in the eluent, then a positive disturbance will appear on the base line corresponding to this higher level of carbonic acid. The disturbance will not appear at the void volume of the system but at a position that is consistent with the retardation of carbonic acid by the hydrogen form of a cation exchanger. By parallel reasoning, an injection that is deficient in carbonate species relative to the prevailing background level will give a negative response that is retarded in exactly the same way as an excess. This is the origin of the carbonate dip.

The cause of its drifting is precisely the same as that causing eluite peak drift. Since carbonic acid is not formed until eluent reaches hydrogen form resin it follows that the magnitude of the retardation will depend on the ''age'' of the suppressor column. Thus the dip will be most retarded with a fresh suppressor and will gradually drift to earlier elution times as this bed exhausts.

There will usually be some point in the process of suppressor exhaustion where the carbonate dip co-elutes with an eluite, subtracting from its response and thereby introducing a source of potentially serious error in the determination of that eluite. This phenomenon provides further argument for employing small volume suppressors and was another stimulus for developing membrane suppressors.

Theoretically, negative disturbances of this sort should not occur with hydroxide eluents. The fact that they do is usually indicative of impurities in the sodium hydroxide being used, the most likely being carbonate.

7.3.4d. Preexhaustion Drop in Conductance in Carbonate Systems

Users of carbonate eluents and column suppressors may have been perplexed, as we were for some time, by the drop in effluent conductance that occurs just before a hydrogen form resin exhausts. One would expect the conductance to rise quite rapidly as the weakly conducting carbonic acid is replaced by the highly conducting sodium carbonate–bicarbonate mixture. This in fact occurs, but not before the conductance dips briefly. For us, the effect at first evoked more curiosity than it did concern, but we noticed it again in another connection where the consequences were more serious and provided added motivation for us to seek an explanation for it (Chapter 5, Section 5.3).

The basis of the effect is revealed in the calculation of Appendix C but may be stated as follows. At the onset of exhaustion, the first traces of sodium that appear in the suppressor effluent must be accompanied by an equivalent level of bicarbonate ions, which in turn depresses the dissociation of carbonic acid in keeping with its dissociation equilibrium. The increase in conductance contributed by the sodium ions is more than offset by the loss of the much more mobile hydronium ions and the conductance drops; if the conductivity meter is at a particularly high sensitivity setting then this drop can appear to be quite dramatic. As neutralization of the carbonic acid proceeds and the sodium rises, this relatively minor effect is eventually overwhelmed by the rising concentration of sodium and bicarbonate species and the conductance then rises rapidly as expected (Figure 7.9).

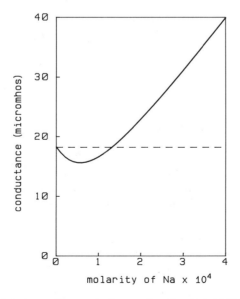

FIG. 7.9. The calculated specific conductance of carbonic acid (0.005 M) in the early stages of its neutralization by sodium hydroxide.

7.3.5. Problems with Column Suppressors

While column-based suppressors were employed in the earliest commercial IC instruments and many applications developed around their use, they suffered from a number of drawbacks, which may be summarized as follows:

1. They required regeneration or replacement. Replacement of suppressor beds was never seriously considered as a convenient or cost effective way of coping with their exhaustion, so they were invariably regenerated. Regenerants were applied as relatively concentrated solutions (one molar sulfuric acid or sodium hydroxide usually, depending on the mode of analysis), so little pumping time was needed to supply the requisite amount of regenerant ion. The interruption for regeneration was prolonged, however, by the often lengthy water rinsing that was necessary to bring the background conductance to an acceptably low and steady level.

2. Peak drifting and the "carbonate dip" were bothersome features.

3. The response of some species was reduced, apparently either by reaction with the acid form of the hydrogen form resin or through reactions catalyzed by it. Nitrite ion was susceptible to this kind of difficulty and would give a response that appeared to depend on the amount of contact with hydrogen form resin. Thus it would be more affected the larger the volume of the suppressor and its response would rise as a suppressor exhausted.

4. The void volume of the suppressor bed was an additional source of extraseparator bandspreading. In the early days of IC this effect was not a serious drawback, but as more efficient columns were developed and the need for more rapid separation grew, this extra source of bandspreading became of more legitimate concern.

There were sufficient grounds, therefore, for developing an alternative means of suppression that would remove or at least minimize these undesirable effects.

7.3.6. Membrane Suppressors

The first anion separation by suppressed IC, in 1971, used a membrane suppressor.[15] Prior to this time, sulfonated lengths of polyethylene surgical tubing had been used to study the possibility of Donnan dialysis concentration of magnesium from seawater. The devices were still around when the first anion separation on a surface agglomerated resin was attempted, and it seemed a logical means of removing the sodium hydroxide eluent. The proposal was to direct the effluent from the separator column through the lumen of the tubular membrane while the outside was in contact with a source of hydronium ions. Thus sodium ions would cross the membrane and be replaced by an equivalent number of hydrogen ions, thereby accomplishing the desired end—suppression of sodium hydroxide. To eliminate leakage of regenerant into the lumen, instead of using a solution of an acid on the outside, the coiled tube was immersed in a stirred bath of small particle size (200–400 U.S. seive size) Dowex 50W × 8 in the hydrogen form. This perfectly nondiffusible acid provided an adequate supply of hydronium ions by "bumping" contacts with the external surface of the sulfonated tube. This first membrane suppressor was capable of suppressing about 0.004 meq of sodium hydroxide per minute. While the membrane device coped with the majority of the sodium hydroxide, the low level that it passed was removed by a small "polishing" column of hydrogen form resin. Although this arrangement performed well, the membranes were brittle and subject to burst under mild excess pressure. Furthermore, column suppressors proved to be much more robust and were so easily prepared that the membrane approach was set aside in the early, rapid development of suppressed IC.

In the late 1970s, when the drawbacks of column suppressors were becoming more acute, the tubular membrane concept was reexamined and developed to provide much more sturdy and practical devices.

The first of these new devices, developed by Stevens et al.,[16] employed a bundle of sulfonated polyethylene fibers (approximately 300 μm i.d., 380 μm o.d.) enclosed in a tube and shell arrangement for supplying a counterflowing stream of 0.02 N sulfuric acid as regenerant. A schematic illustration of the hollow fiber suppressor is provided in Figure 7.10.

These devices added somewhat more bandspreading than an average column suppressor but they did eliminate most of the other problems. Thus Stevens was able to demonstrate that while the peak height response of acetate ion using a suppressor column varied from 29.8 to 76 units owing to peak drifting, a hollow fiber device, on the other hand, gave an essentially flat response of about 44 units during a similar period of usage.

The use of a membrane device cannot eliminate the carbonate dip, but it does ensure that it remains in the same place. In these first hollow fiber devices it merged with the water dip and tended to co-elute with fluoride ion.

In a later development Stevens et al.[17] used a single fiber of Nafion,* a commercially available sulfonated perfluorinated membrane, into which they introduced small spheres of styrene-divinylbenzene copolymer to essentially form a necklace of beads within the lumen. This had two beneficial effects: it reduced the void volume and hence bandspreading contributed by the suppressor and it improved mass transfer within the lumen so that shorter lengths of device could supply the requisite amount of suppression. Figure 7.11 shows a comparison of packed and unpacked hollow fiber suppressors, and while the shape of later eluting peaks is similar in both devices, as would be expected, the improvement in efficiency of the packed device is very evident for ions that elute early. When compared to a typical packed bed suppressor the packed fiber device gave somewhat less bandspreading.

*Nafion is a trademark of the Du Pont Company.

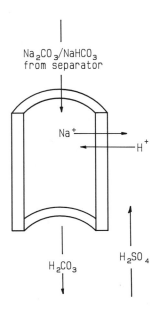

FIG. 7.10. How a hollow fiber membrane suppressor works. Hydronium ions from sulfuric acid flowing on the outside of the cation exchange membrane exchange for sodium in the sodium bicarbonate/sodium carbonate eluent flowing in the opposite direction through the fiber lumen. The eluent is converted to carbonic acid continuously.

Hanaoka *et al.*, working independently, described a very similar type of suppressor device.[18]

Dasgupta and co-workers have made significant contributions to the development and understanding of membrane suppressors. They developed filament-filled, helically wound hollow fiber configurations and demonstrated that they contributed less to bandspreading then bead-filled fibers. Their research also did much to clarify the mass transfer characteristics and requirements of these membrane systems.[19–22]

The net effect of these developments was to eliminate most of the drawbacks of the column suppressors while not adding anything more to bandspreading than the columns had already. As a result, column suppressors were gradually phased out of commercial suppressed IC instruments to be replaced by both anion and cation versions of the packed hollow fiber design.

In late 1985 Dionex Corporation introduced another significant improvement in the development of suppressors.[12] This too was a membrane device operating under the same principles as the original, but with markedly improved suppression capacity and reduction in void volume. It employed a dual-membrane design based on very thin membranes of proprietary composition. The construction of the Dionex "MicroMembrane Suppressor"* is schematically illustrated in Figure 7.12. The effluent from the separator flows between two 50-μm-thick cation exchange membranes (anion analyzer) that are held apart by

*MicroMembrane Suppressor is a trademark of Dionex Corporation.

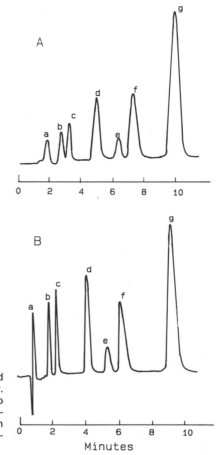

FIG. 7.11. Comparison of (A) an unpacked and (B) a packed hollow fiber suppressor. The peaks a,b,c, etc. are due, respectively, to fluoride, chloride, nitrite, phosphate, bromide, nitrate, and sulfate. (From Ref. 17 with permission. Copyright 1982 American Chemical Society.)

an intermediate screen that is lightly sulfonated. Regenerant acid is pumped countercurrently and externally to the membranes in small volume channels that are also partially filled with plastic screens. The screens appear to fulfill the same functions as the bead or filament packing of fibers; they reduce the void volume in the effluent channel and improve mass transfer rates in both the interior and exterior compartments.

The void volume of this device is very low—typically around 50 μl—and it has a remarkable ability to exchange high levels of influent ions. For example, an anion suppressor is capable of removing the sodium from 0.1 N sodium hydroxide when it is applied at a flowrate of about 2 ml/min. The thinness of the ion exchange membranes aggravates somewhat the problem of regenerant diffusion

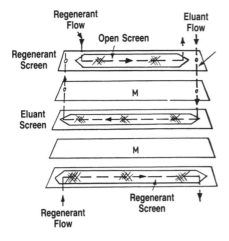

FIG. 7.12. Elements of Dionex Micro-Membrane Suppressor. Eluent flows between two flat membranes (M)—cation exchange membranes for anion analysis, anion exchange membrane for cation analysis. Regenerant flows in the opposite direction through compartments on the outside of the membranes. Plastic screens placed in the eluent and regenerant compartments promote efficient mass transfer between the solution and membrane phases. (Courtesy Dionex Corp.)

into the middle compartment, but this can be reduced by using regenerant acids of relatively high molecular weight.

Maintaining a supply of regenerant solution has always been part of the routine of suppressed IC whether using column or membrane-based suppressors. A recent development of the Dionex Corporation dispenses with the frequent need to replenish regenerant. It employs a very large cartridge of cation exchange resin (anion analysis) connected to the regenerant compartment(s) of a Micro-Membrane suppressor by way of a small circulating pump (Figure 7.13). Acid at low concentration fills the void volume of the cartridge and the regenerant compartment and performs the usual and vital function of supplying hydronium ions to the interior compartment. However, the spent regenerant containing sodium, instead of going to waste, is passed to the cartridge, where sodium is replaced by hydronium and the cartridge effluent is returned to the suppressor. The ion exchange capacity of the 1-l cartridge of high-capacity resin (about two equivalents) is adequate for several months of use on a typical busy anion analyzer. An analogous device is available for cation analysis.

Thus research in suppressed IC has eventually found a way of tapping the major benefit of a massive suppressor—its impressive capacity to suppress eluent—without the concomitant penalties of large void volume and dead times that would result from connecting directly to it.

7.3.7. Removal of CO_2 from Column Effluent

Stripping of carbon dioxide from the typical effluent of suppressed IC as a means of further reducing background conductance has been studied by Sunden and co-workers.[23] They use a postsuppression device whose key element is a length of gas permeable poly(tetrafluoroethylene) tubing through which the

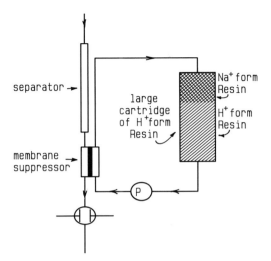

FIG. 7.13. Arrangement for the automatic regeneration of spent regenerant from a membrane suppressor (anion analysis). P is a regenerant circulating pump. The heavily shaded area in the membrane suppressor represents the cation exchange membrane that separates the eluent and regenerant compartments.

effluent from the suppressor flows before passing it to the conductivity cell. The outside of the tubing is kept at a low pressure and the carbon dioxide diffuses across to the low-pressure side. As much as 90% reduction of the background conductance is claimed by this method. The extra volume of the carbon dioxide stripper does degrade eluite peak sharpness somewhat.

7.3.8. Nonlinearity Effects in Suppressed Conductometric Systems

In some suppressed conductometric systems there is a nonlinear dependence of peak height on sample amount. This is common when the eluites produce either weak acids or weak bases in the suppressor and is caused by the decrease in eluite dissociation with increasing concentration. Bouyoucos[14] was one of the first to report the effect and to suggest means of correcting for it.

Nonlinearity can also occur when the eluites produce strong electrolytes in the suppressor, and a number of workers have reported on it and discussed its origin.[24–26] It is observed, for example, with carbonate eluents and is a direct result of the tendency of the strong acid eluite to suppress the dissociation of the cabonic acid in which it is present. It is a relatively simple matter to calculate the conductance of a solution of a strong acid in a weaker acid; the results for solutions of hydrochloric acid in carbonic acid are shown in Figures 7.14 and 7.15. For comparison, the conductances of hydrochloric acid in water are also

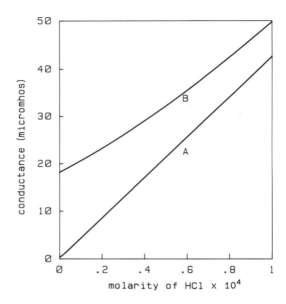

FIG. 7.14. Specific conductance (calculated) of hydrochloric acid in A, water and B, carbonic acid (0.005 *M*) for very low concentrations of hydrochloric acid.

presented. It is evident that the conductance of solutions of HCl in carbonic acid is not linear with HCl concentration at low concentrations (Figure 7.14) but becomes so at higher concentrations (Figure 7.15). The explanation of these effects is as follows.

When hydrochloric acid is added to carbonic acid it suppresses the dissociation of carbonic acid, so the net conductance is not equal to the added conductances of the separate acids but is instead somewhat less than this. As the concentration of the strong acid is increased, a point is reached when essentially all of the carbonic acid exists as its associated nonconducting form and any further additions of HCl will produce an effect that is proportional to the amount of HCl added, in other words a linear response. Theoretically, this effect is also present when water is the background, but since water is such an extremely weak "acid" the amount it contributes to the conductance is so minute by comparison that any nonlinearity that exists is imperceptible.

Putting this effect in chromatographic terms the base-line conductance of the carbonic acid is analogous to a "spongy platform" on which the HCl peak rests. The HCl compresses this platform to a degree that depends on the concentration of the HCl in the peak. Eventually, if the eluite peak is concentrated enough, the platform is totally compressed and any further increases in concentration produce no further compression. Besides the nonlinearity, another effect of this compressible base line is that the peak due to a given amount of HCl

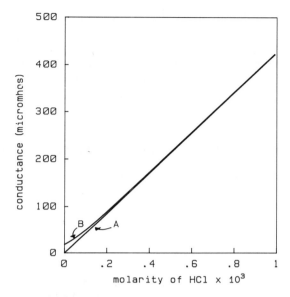

FIG. 7.15. Specific conductance (calculated) of hydrochloric acid in A, water and B, carbonic acid (0.005 *M*) over a wider range of hydrochloric acid concentration than Figure 7.14.

will not appear as tall in a carbonic acid background as it would in water, all other factors being the same.

The mathematical basis of these various effects is given in Appendix D.

7.4. CONDUCTOMETRIC DETECTION: NONSUPPRESSED

The separation and conductometric detection of ions without the use of a suppressor device was first reported in 1979.[4,5] Subsequently, Fritz and co-workers published a series of papers and a monograph[27] describing the new method and prescribing conditions for its application to the analysis of both cation and anion mixtures. More or less concurrent with this and other research there has been the commercialization of instrumentation embodying the techniques.

The use of conductometric detection in this way appears under a variety of names. Single-column ion chromatography (SCIC) was an early name that distinguished it from the dual-column methods of early IC, but one that nowadays can be confusing in light of the many other IC methods that use a single column, and the fact that membrane devices have replaced the second column of the original IC method. "Direct conductometric detection" has some validity since the eluent is passed directly to the detector without first going through a suppressor; on the other hand Dasgupta[28] argues that the name "indirect conductometric detection" is also justified. The name "nonsuppressed conductometric detection" has been employed in this book; it is commonly used elsewhere, it leaves less room for ambiguities than other terminology, and it brings out the principal distinction that separates the two major conductometric techniques, at least from an operational standpoint.

7.4.1. The Basis of Nonsuppressed Conductometric Detection

Nonsuppressed approaches to conductometric detection confront the same problem as their suppressed counterparts, namely, the measurement of low levels of eluite in an environment that is rich in the property being monitored. The problem may be elaborated by considering the conductometric detection of an eluite ion eluted from an ion exchanger. In the example an anion exchange procedure is assumed in which the eluent is the sodium salt of a monovalent displacing ion denoted simply E^-. After equilibration with Na^+E^- the column effluent from the anion exchange column will be identical to the eluent and its conductivity response, denoted L_1, will be equal to $1000C_E(\lambda_{Na} + \lambda_E)$ μmho, where C_E is the concentration of the eluent and λ_{Na} and λ_E are the limiting equivalent conductances of the sodium and eluent ions, respectively. The following additional assumptions are implicit in this expression:

1. The Kohlrausch law of independent mobilities applies. This is legitimate in view of the low concentrations (usually $< 10^{-3}\ M$) that are common to IC. The use of limiting conductances is justified on the same grounds.
2. The cell constant is unity.

Since the concentration of sodium co-ions remains fixed at the eluent concentration, the sum of the concentrations of eluite and eluent anions must likewise equal C_E so as to obey electroneutrality demands. An injection of eluite denoted S^- will eventually appear at the column exit giving rise to a concomitant change in the concentration of the displacing ion (Figure 7.16). Consequently, if the concentration of eluite at its peak is C_S it follows that the concentration of E^- at that same point is $C_E - C_S$. Therefore the conductance at the peak of S^-, denoted L_2, is the sum of the contribution of all the ions present at that point and is given as

$$L_2 = 1000[C_E\lambda_{Na} + (C_E - C_S)\lambda_E + C_S\lambda_S] \tag{7.21}$$

It follows that the change in conductance at the peak, ΔL, is then

$$1000C_S(\lambda_S - \lambda_E) \tag{7.22}$$

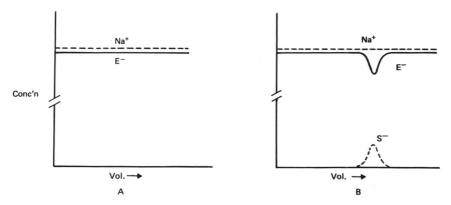

FIG. 7.16. Ion exchange elution of analyte (anion case). Shows how the elution of analyte anion (S^-) causes a concerted and equivalent "dip" in the concentration of the eluent anion (E^-).

The magnitude of the response is therefore proportional to the difference in conductance between the analyte and eluent ions as well as to the concentration of the eluite. Obtaining any response is clearly dependent on there being a difference between the equivalent conductances of the eluite and eluent ions. Obtaining a detectable response, as we have seen (Chapter 6), depends on the magnitude of ΔL compared to the average noise contributed by the background eluent. This will be pursued in a later section.

7.4.2. Eluent Choice

The fact that this noise is directly proportional to the concentration level of eluent means that the detection of very small amounts of eluite are facilitated by using dilute eluents. This in turn precludes the use of commonly available ion exchangers of high capacity since they require very high concentrations of eluent to displace eluites with requisite speed (Chapter 4, Section 4.2.5). Fritz *et al.* developed a variety of low-capacity resins to match the eluting power of the dilute eluents; the capacity of their early resins was in the range from 0.007 to about 0.1 meq/g.

They also noted that the requirements of low eluent concentration and adequate displacing power could be more easily met if they used potent displacing ions. Organic acid anions such as benzoate, phthalate, and sulfobenzoate had this property and had conductances that differed significantly from those of the common inorganic eluite ions that were the focus of their interest.

Choosing among various eluents was influenced principally by the affinities of the eluite species, benzoate being preferred for mixtures such as acetate, bicarbonate, fluoride, chloride, nitrite, and nitrate. Divalent eluites and other intractable ions such as thiocyanate, perchlorate, or iodide were not effectively eluted by benzoate and required that phthalate or sulfobenzoate be used.

Analogous schemes were developed for cation separations.[27] Here there was considerable overlap with suppressed techniques in that they used similar low-capacity cation exchange resins in conjunction with dilute hydrochloric or nitric acids for alkali metals and ammonium, and divalent displacers such as diprotonated form of ethylene diamine to elute more persistent species such as ions of the alkaline earth series.

The treatment of conductance and conductance changes (see above) applies analogously to cation analysis systems. Notable in using acid eluents is the appearance of eluites as negative peaks or troughs since in this case the conductance of the eluent ion, hydronium, is much greater than that of any other cation. Negative peaks are also characteristic of hydroxide eluents in anion analysis. Since it is chromatographic convention to display signals in an upscale fashion, regardless of whether or not the signal is increasing, it is common practice to reverse the polarity of the output signal when negative responses are indicated.

As an alternative to common salts of weak organic acids, Gjerde and Fritz[29] proposed dilute solutions of the acids themselves. They observed that analyte anions elicited a greater response in an acid eluent than in the salt of the same acid and on this basis suggested that in some cases, notably benzoic acid, the acid eluent would be a better eluent from a detection standpoint. The basis of this conclusion came from their definition of sensitivity[30] that equated sensitivity and response, that is "change in detector signal per unit concentration." Because response alone is an inadequate measure of the ability of a system to detect low levels of an analyte, that is, its sensitivity in the truest sense, a proper comparison of sensitivity in the various eluents is not directly accessible from this work since no measurements of noise were reported.

7.5. A COMPARISON OF SUPPRESSED AND NONSUPPRESSED CONDUCTOMETRIC DETECTION

With this background on conductometric detection we are at a point where it is appropriate to essay a comparison of the two approaches.

There are some obvious ways in which the two conductometric methods are alike and in which they differ. For example, both exploit electrical conductance to measure analytes and employ the same transducer, the conductivity cell. The most obvious difference is that one method uses a suppressor device and the other does not. There are other distinctions, however, that are less obvious but are nonetheless very significant. We will consider two of them.

Two very important measures of efficacy in any chromatographic method are sensitivity and dynamic range, and it is in these respects that the suppressed and nonsuppressed approaches are notably different. To explain these fundamental distinctions we will employ simulations of chromatographic experiments that enable a side-by-side comparison of the two conductometric methods. Four points should be considered in this comparison:

1. On what does sensitivity depend?
2. What factors limit sensitivity in the two methods?
3. What measures improve sensitivity?
4. How do these measures affect dynamic range?

7.5.1. Simulated Experiments Using Suppressed and Nonsuppressed Methods

Of the several yardsticks by which the two conductometric methods may be compared, sensitivity is the one that seems to elicit the most misunderstanding. No systematic experimental study of relative sensitivities has been reported as yet, but, because of the good basic knowledge of electrolyte conductance and of the ion exchange process it is possible to treat the matter effectively from a theoretical point of view. The following is such a treatment augmented by a computer simulation of the two conductometric techniques as applied to the IC analysis of cations.

The cation example is chosen because of the great degree of overlap in the two methods as commonly practiced; they both use a low-capacity cation exchanger as stationary phase and a dilute mineral acid such as hydrochloric or nitric acid as mobile phase.

In most of the hypothetical examples the following parameters are assumed fixed with values as noted: volume of separator bed, 2 ml; specific capacity of separator, 0.01 meq/ml; selectivity coefficients for sodium and potassium ions relative to the hydronium, 1.6 and 2.3, respectively; volume of injected sample, 0.1 ml.

From these values and knowing the concentration of acid eluent it is possible to calculate the elution volumes of the eluites in the usual way (4.22). The void volume of a 2-ml bed is assumed to be 0.8 ml.

To calculate the profiles of the eluting bands, symmetrical Gaussian peaks and a total column efficiency of 3000 plates are assumed. The membrane suppressor is assumed to be of small void volume and therefore to subtract negligibly from this efficiency.

In the case of nonsuppressed detection the effluent from the separator is passed directly to a conductivity cell–meter combination, while for suppressed it first passes through a membrane suppressor and then to the conductance detector.

In calculating conductances, Kolrausch's law of independent ion mobilities is assumed to apply, so conductances may be calculated by summing the conductances of the separate cations and anions in the system [see equation (7.21), for example]. Limiting equivalent conductances of the various ions are assumed to apply (see Appendix B). A cell constant of unity is assumed.

The simulation ignores the conductance disturbance at the void since it is not relevant to the objectives of the exercise.

In the nonsuppressed mode the eluites are revealed by the decrements in conductance that they cause when the hydronium ions are replaced by the less conducting sodium and potassium ions:

$$\Delta L \text{ for sodium } = 1000C_{Na}(\lambda_{Na} - \lambda_H) \qquad (7.23)$$

where C_{Na} is the concentration of sodium ion at any point in the sodium band.

In the chromatograms of the nonsuppressed mode, it will be noted that conductance values decrease in the positive direction of the ordinate. This simulates the common practice of reversing the signal polarity when negative peaks are obtained.

In the suppressed mode the eluites appear as bands of sodium and potassium hydroxide in essentially deionized water. The positive deflection in conductance caused by the sodium band, for example, is calculated from the formula

$$\Delta L = 1000C_{Na}(\lambda_{Na} + \lambda_{OH}) \qquad (7.24)$$

In a hypothetical suppressed mode one could in principle assume the extremely low conductivity of pure water as a base line, but a real system is more fairly represented by choosing a value of about 2 μmho. This takes account of the small amount of impurities that are inevitable in the eluent reagent and any slight leakage from the suppressor.

It remains to add the important contribution of noise. Various sources of noise are possible, but a reasonable and the most likely one is imperfect control of the temperature of the effluent reaching the conductivity cell. It is assumed that the tolerance in the temperature control is $\pm 0.005°C$. The temperature coefficient of conductance is around 2%/°C so that the uncertainty in conductance is therefore $\pm 0.01\%$.

The intrinsic noise of the detector is assumed to be 0.01 μmho and is the noise contributed at all times regardless of the signal arising from the background. This level of noise will become important only when the background signal is low.

The graphical representation of the results as well as their calculation is handled by a computer. The random feature of noise is simulated by employing the computer's ability to generate a random factor that can be "patched on" to the calculated signals.

The main purpose of the hypothetical experiment is to examine the detectability of eluites by the two methods. For all but one of the experiments the concentration of eluent is fixed at 0.005 M. This level of acid in combination with the other parameters yields good peak separation within the acceptable time of less than 20 min.

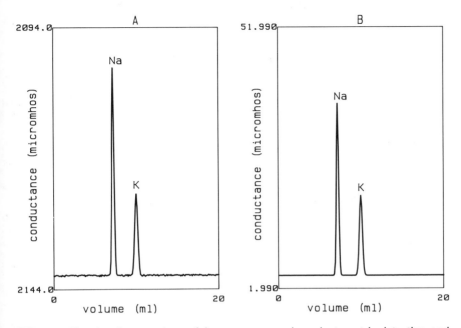

FIG. 7.17. Simulated comparison of A, nonsuppressed conductometric detection and B, suppressed conductometric detection. Separation of sodium and potassium on a low-capacity cation exchanger using hydrochloric acid as eluent. In this experiment the concentrations of sodium and potassium in the sample were assumed to be 10 ppm in each ion. The chart span was 50 μmho full scale. For other conditions see text.

In the first experiment (Figure 7.17) each device receives an injection wherein the concentration of the sodium and potassium are each 10 parts per million. In the nonsuppressed mode the base-line conductance is roughly 2100 μmho so the noise due to temperature instability is approximately 0.2 μmho. For the suppressed mode, because of the much lower background conductance, the noise that is relevant is the intrinsic noise of 0.01 μmho. Comparison of the two methods shows that while noise is imperceptible in the suppressed mode it is barely perceptible in the nonsuppressed, and therefore, in being able to precisely quantify the two species by peak height measurement, there is little to choose between the two methods.

However, as the eluite concentration is lowered to 1 ppm the problem of noise is clearly beginning to affect the accuracy of measurement in the nonsuppressed mode (Figure 7.18). When the concentration is lowered still further to 0.1 ppm the noise has overwhelmed the signals in the nonsuppressed mode while the suppressed system is by comparison unafflicted by noise (Figures 7.18 and 7.19).

The great difference in noise between the two methods is of course caused

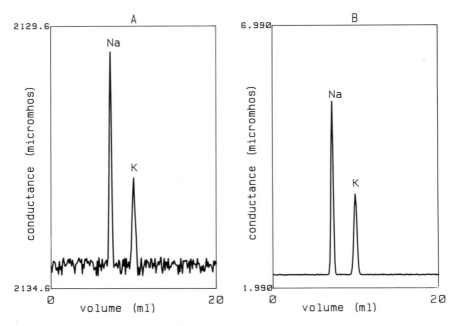

FIG. 7.18. Same chromatographic conditions as the simulation represented in Figure 7.17 but sample concentrations changed to 1 ppm in each of the analyte cations. Chart span was 5 μmho full scale.

by the extreme difference in base-line signal, about 2100 μmho in the nonsuppressed, 2 μmho in the suppressed mode. It is clear, therefore, that lowering the concentration of eluent will lower the background conductance and hence the noise in the nonsuppressed mode, but in order that the elution still take place in approximately the same time, the capacity of the ion exchange packing must be lowered proportionally.

The results of such an adjustment are shown in Figure 7.20. The concentration of HCl has been decreased from 0.005 M to 0.005 M and the resin capacity from 0.01 to 0.001 meq/g. The improvement in detectability is apparent and the experiment illustrates the rationale for using resins of lower capacity, a common strategy for improving detectability in the nonsuppressed mode.

This measure does, however, involve a compromise since lowering the eluent strength and the resin capacity impairs the system's ability to resist overload. In contrast, the suppressed system, with a tenfold greater resin capacity, has a tenfold advantage over the nonsuppressed system in overload capacity. This will be reflected in a greater dynamic range for the suppressed method.

It is appropriate at this point to mention that while the suppressed method also uses low-capacity resins and dilute eluents it does so for a quite different reason, namely, to reduce the load on the suppressor. Modern suppressors, with

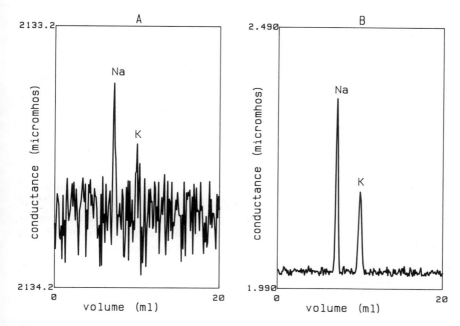

FIG. 7.19. Same chromatographic conditions as the simulation represented in Figure 7.17 but sample concentrations changed to 0.1 ppm in each of the analyte cations. Chart span 1 μmho full scale in the nonsuppressed experiment (A) and 0.5 μmho in the suppressed experiment (B).

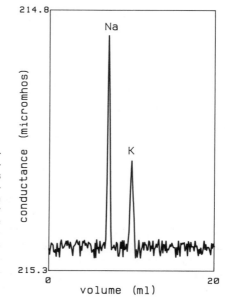

FIG. 7.20. The effect of lowering the eluent strength on detectability in nonsuppressed conductometric detection. In this simulation the resin capacity and the hydrochloric acid concentration have both been reduced to a tenth of what they were in the previous experiments. The concentrations of sodium and potassium in the sample were 0.1 ppm in each. This is the simulated chromatogram using nonsuppressed detection illustrating the beneficial effect on detectability of lowering the eluent strength.

their much higher suppressing capacity, impose less of a constraint on the concentration of eluent that may be used and in effect expand the dynamic range of conductometric detection.

The ability to handle a wide range of analyte levels is one of the ways in which analytical techniques are evaluated. This exercise demonstrates the advantages of the suppressed IC method in this regard. It also shows how the dynamic range of a method can be dictated by the chemistry of the separation and not by any limitations in the transducer.

The series of simulations also serves to underline the importance of exploring several properties of conductometric IC systems when comparing their relative merits.

The exercise also reveals the pitfall of using too restricted a definition of sensitivity. If one compares the peak heights of the analytes in Figure 7.17 it is apparent that the responses in the nonsuppressed mode are in fact somewhat greater than in the suppressed mode. It is erroneous, however, to conclude that the sensitivity of the former is greater than the latter because of this greater response. When noise is considered alongside response, as it properly should be and is in Figure 7.19, then the conclusion that eluent suppression leads to superior sensitivity is the correct one.

Figure 7.17 also illustrates how, *under the proper circumstances,* comparing responses can be a legitimate means of comparing sensitivities. By comparing the heights of the sodium and potassium peaks it is evident that either method is more sensitive for sodium. Comparisons of this sort have validity since they are made within the same method *with the same level of noise applying.*

Finally it should be pointed out that these simulations are designed to compare the two conductometric approaches under identical detection conditions and to demonstrate the beneficial effect of using a suppressor, *all other factors being the same.* They do not indicate the absolute capabilities of either method with respect to detectability limits.

7.6. GRADIENT ELUTION AND CONDUCTOMETRIC DETECTION

Frequently an ion chromatographic separation involves species of widely different affinities for the stationary phase. In such a case, eluent conditions that favor the resolution of the most weakly held species are often unsuitable for the more tightly held ions in that they lead to overly long elution times, and using a strong eluent, while it may elute the strongly held ions in a reasonable time, usually destroys the resolution of the early eluting species. Gradient elution is one means of solving the problem. In gradient elution, displacement begins with an eluent of low displacing power and progresses either gradually or in a series of steps to an eluent of greater potency. For many detection methods this poses little problem. If the detector is sensitive only to the eluites and not to the components of the eluent then it is possible to make quite large changes in eluent composition without affecting the detectability of the eluites. The application of gradient elution in HPLC with UV detection is a good example of this.

When eluite and eluent ions share the property being monitored (bulk property detectors) then signal due to eluite can be lost in the large change in base-line signal that accompanies the gradient. Nonsuppressed conductometric detection suffers from this problem, although a recently described strategy relieves it to some extent.[31] In this so-called "isoconductance gradient" approach the conductivity of the weaker eluent is boosted by choosing a co-ion from among those with higher conductance while the higher conductivity of the stronger eluent is subdued by employing co-ions of low conductance. While it is possible to essentially flatten out the conductance gradient by this technique, two problems remain. In the first place, the limited range of ion conductances enforces a limit on the range of eluent ion concentration that can be used in the gradient. As a result, the degree of compression of the k-prime range by the gradient is limited. Secondly, since this is a nonsuppressed conductometric method it has the sensitivity limitations of that approach.

Suppressed conductometric methods, on the other hand, since they essentially eliminate the eluent, are more amenable to gradient adaptations. Some early attempts[32−34] were less than successful, but with the advent of newer membrane devices with their high suppression capacity, gradient elution can now be applied with great effect to suppressed systems.[35,36] In this regard the enhanced status of sodium hydroxide as an eluent in anion analysis and particularly in gradient elution is especially noteworthy. Sodium hydroxide is from a suppression standpoint the ideal eluent since its product is water, but its low ion exchange affinity has limited its usefulness as a displacing ion. Low affinity can be counteracted to some extent by employing higher concentrations of eluent but this measure was limited in the past by the low capacity of the available suppressor devices. The ability of modern suppressors to handle relatively high concentrations of sodium hydroxide is therefore a boon to anion analysis by gradient elution.

While it is an effective way to handle ions of widely diverse affinities, gradient IC is not without its problems. For example, impurities in eluent reagents can build up on the separator bed in the early stages of the elution and be released later as the separator is swept by the more concentrated eluent. Rocklin *et al.* have examined this problem in some depth and suggest measures to alleviate it.[35] The problem should diminish with the availability of purer reagents.

7.7. CONDUCTOMETRIC DETECTION IN ION CHROMATOGRAPHY—A SUMMARY

Conductometric detection by either suppressed or nonsuppressed techniques offers chromatographers a versatile means of quantifying the eluites in ion exchange and, as will be shown in other parts of the book, in closely related modes of separation. By making a great variety of ions accessible to conductometric detection the field of chromatography has been opened up to ions without chromophores.

The complexities added by the detector hardware are comparable in the two methods. The criticisms of the older column-based suppressors are no longer relevant with the advent of simple, low-volume, continuously regenerated suppressors. The requirement of maintaining very low tolerances of temperature variation in the nonsuppressed mode and the added expense and complexity that rigorous thermostating involves is frequently omitted from comparisons of the two methods.

While there are many ways in which the methods have been claimed to differ from an operational standpoint they are also difficult to verify or quantify. Stability of base lines, the time it takes a system to be operational and stabilized from a "cold start," and susceptibility to disturbances such as "pseudo" and system peaks, are some of the issues that the user should explore in any total evaluation of conductometric IC.

It is on the question of sensitivity that there is most debate. One still frequently encounters the claim that some new development "eliminates the need for suppression." There will always be a need for superior detection methods as analysts pursue lower and lower limits of detectability. Suppression provides that extra measure of sensitivity while maintaining a high level of overload capacity.

REFERENCES

1. H. Small, T. S. Stevens, and W. C. Bauman, Novel Ion Exchange Chromatographic Method Using Conductimetric Detection, *Anal. Chem.* **47,** 1801–1809 (1975).
2. H. Small and W. C. Bauman, Apparatus and Method for Quantitative Analysis of Ionic Species by Liquid Chromatography, U.S. Patent No. 3,920,397 (1975).
3. H. Small and T. S. Stevens, Chromatographic Analysis of Ionic Species, U.S. Patent No. 3,925,019 (1975).
4. K. Harrison and D. Burge, Anion Analysis by HPLC, Pittsburgh Conference on Analytical Chemistry, Abstract No. 301 (1979).
5. D. T. Gjerde, J. S. Fritz, and G. Schmuckler, Anion Chromatography with Low-Conductivity Eluents, *J. Chromatogr.* **186,** 509–519 (1979).
6. R. A. Robinson and R. H. Stokes, *Electrolyte Solutions,* p. 145, Academic Press, New York (1955).
7. H. H. Willard, L. L. Merritt, J. A. Dean, and F. A. Settle, *Instrumental Methods of Analysis,* 6th. ed., p. 783, Wadsworth Publishing, Belmont, California (1981).
8. J. Weiss, *Handbook of Ion Chromatography,* Dionex Corporation, Sunnyvale, California (1986).
9. T. S. Stevens and H. Small, Surface Sulfonated Styrene Divinylbenzene—Optimization of Performance in Ion Chromatography, *J. Liquid Chromatogr.* **1**(2), 123–132 (1978).
10. H. Small and T. S. Stevens, High Performance Ion Exchange Composition, U.S. Patent, No. 4,101,460 (1978).
11. H. Small and J. Solc, Ion Chromatography—Principles and Applications, in *The Theory and Practice of Ion Exchange* (M. Streat, ed.), The Society of Chemical Industry, London (1976).

12. J. Stillian, An Improved Suppressor for Ion Chromatography, *L. C. Mag.* **3**(a), 802–812 (1985).
13. C. A. Pohl and E. L. Johnson, Ion Chromatography—The State of the Art, *J. Chromatogr. Sci.* **18**, 442–452 (1980).
14. S. Bouyoucos, Determination of Ammonia and Methylamines in Aqueous Solutions by Ion Chromatography, *Anal. Chem.* **49**, 401–403 (1977).
15. H. Small, unpublished results.
16. T. S. Stevens, J. C. Davis, and H. Small, Hollow Fiber Ion-Exchange Suppressor for Ion Chromatography, *Anal. Chem.* **53**, 1488–1492 (1981).
17. T. S. Stevens, G. L. Jewett, and R. A. Bredeweg, Packed Hollow Fiber Suppressors for Ion Chromatography, *Anal. Chem.* **54**, 1206–1208 (1982).
18. Y. Hanaoka, T. Murayama, S. Muramoto, T. Matsuura, and A. Nanba, Ion Chromatography with an Ion-Exchange Membrane Suppressor, *J. Chromatogr.* **239**, 537–548 (1982).
19. P. K. Dasgupta, Linear and Helical Flow in a Perfluorosulfonate Membrane of Annular Geometry as a Continuous Cation Exchanger, *Anal. Chem.* **56**, 96–103 (1984).
20. P. K. Dasgupta, Annular Helical Suppressor for Ion Chromatography, *Anal. Chem.* **56**, 103–105 (1984).
21. P. K. Dasgupta, R. Q. Bligh, and M. A. Mercurio, Dual Membrane Annular Helical Suppressors in Ion Chromatography, *Anal. Chem.* **57**, 484–489 (1985).
22. P. K. Dasgupta, R. Q. Bligh, J. Lee, and V. D'Agostino, Ion Penetration through Ion Exchange Membranes, *Anal. Chem.* **57**, 253–257 (1985).
23. T. Sunden, A. Cedergren, and D. D. Siemer, Carbon Dioxide Permeable Tubing for Postsuppression in Ion Chromatography, *Anal. Chem.* **56**, 1085–1089 (1984).
24. J. Slalina, F. P. Bakker, P. A. Jongejan, L. Van Lamven, and J. J. Mols, Fast Determination of Anions by Computerized Ion Chromatography coupled with Selective Detectors, *Anal. Chim. Acta* **130**, 1–8 (1981).
25. M. J. Van Os, J. Slalina, C. L. De Ligny, W. E. Hammers, and J. Agterdenbos, Determination of Traces of Inorganic Anions by Means of High-Performance Liquid Chromatography on Zipax-SAX Columns, *Anal. Chim. Acta.* **144**, 73–82 (1982).
26. M. Doury-Berthod, P. Giampaoli, H. Pitsch, C. Sella, and C. Poitrenaud, Theoretical Approach of Dual-Column Ion Chromatography, *Anal. Chem.* **57**, 2257–2263 (1985).
27. J. S. Fritz, D. T. Gjerde, and C. Pohlandt, *Ion Chromatography.* Dr. Alfred Huthig, Heidelberg (1982).
28. P. K. Dasgupta, Approaches to Ionic Chromatography, in *Ion Chromatography*, Marcel Dekker, New York (1987).
29. D. T. Gjerde and J. S. Fritz, Sodium and Potassium Benzoate and Benzoic Acid as Eluents for Ion Chromatography, *Anal. Chem.* **53**, 2324–2327 (1981).
30. J. S. Fritz, D. L. DuVal, and R. E. Barron, Organic Acid Eluents for Single-Column Ion Chromatography, *Anal. Chem.* **56**, 1177–1182 (1984).
31. W. R. Jones, P. Jandik, and A. L. Heckenberg, Gradient Elution of Anions in Single Column Ion Chromatography, *Anal. Chem.* **60**, 1977–1978 (1988).
32. T. Sunden, M. Lundgren, and A. Cedergren, Separation of Sulfite, Sulfate and Thiosulfate by Ion Chromatography with Gradient Elution, *Anal. Chem.* **55**, 2–4 (1983).
33. J. G. Tarter, Gradient Elution Ion Chromatographic Determination of Inorganic Anions using a Continuous Gradient, *Anal. Chem.* **56**, 1264–1268 (1984).
34. K. Irgum, Gradient Membrane-Suppressed Anion Chromatography with N-Substituted Aminoalkylsulfonic Acid Salts as Eluents, *Anal. Chem.* **59**, 363–366 (1987).
35. R. D. Rocklin, C. A. Pohl, and J. A. Schibler, Gradient Elution in Ion Chromatography, *J. Chromatogr.* **411**, 107–119 (1987).
36. H. Shintani and P. K. Dasgupta, Gradient Anion Chromatography with Hydroxide and Carbonate Eluents Using Simultaneous Conductivity and pH Detection, *Anal. Chem.* **59**, 802–808 (1987).

Chapter 8

Other Modes of Detection

8.1. INTRODUCTION

While conductometric detection is a widely used and important part of modern IC it is a relative newcomer to the area of chromatographic detection. A number of other detection methods (e.g., photometric, electrochemical) have been extensively employed and developed for HPLC, and, given their considerable merits, it is not surprising that they are applicable to the chromatography of ionic species. In fact, photometric detection coupled to ion exchange or ion interaction separation was a significant part of HPLC practice before modern conductometric methods came on the scene. This chapter will describe the more important of these other methods, indicate how they may be applied to the separation of ions, and point out some of their strengths and limitations in the context in which they are used.

8.2. PHOTOMETERS AND SPECTROPHOTOMETERS

8.2.1. Principles of Photometric Detection

Photometers and spectrophotometers are the most widely used detectors in modern HPLC. They exploit the fact that many species absorb UV or visible radiation and the principle that the absorption is governed by Beer's law, which states that

$$I_T = I_0 e^{-\epsilon l c} \tag{8.1}$$

where I_0 is the intensity of the incident light, I_T is the intensity of the light transmitted through the examined sample, l is the pathlength of the light through the sample, c is the concentration of the light-absorbing species, and ϵ is the molar extinction coefficient of the species at the wavelength of the incident radiation.

The quantity I_0/I_T is a measure of light absorbance and its logarithm is defined as the absorbance, A, of the solution:

$$\log(I_0/I_T) = A = \epsilon l c \tag{8.2}$$

For a fixed pathlength and monitoring wavelength the absorbance is therefore proportional to the concentration of absorbing species and photometers are usually designed to take advantage of this convenient linearity. Photometric detectors will thus include a logarithmic converter so that the output signal is an expression of absorbance. It should be pointed out that deviations from Beer's law are possible and indeed fairly common but in the concentration range in which we will be most interested the linearity is fairly reliable. It does underline again, however, the importance of calibration measurements in chromatography and the hazard in a too liberal use of linear extrapolations.

Figure 8.1 is a schematic showing the essential features of photometric measurement. At the center of the system is the photometer cell where a small segment of solution is examined. An optically flat window admits a collimated light beam; another allows the transmitted light to pass and impinge on a photodetector such as a simple photocell or a photomultiplier.

The light may come from a variety of sources depending on the application. If the sample is to be examined only in the visible part of the spectrum then a tungsten lamp will do, but if UV absorption is to be measured then sources with significant output in the region of 190–400 nm will be necessary; low or medium pressure mercury lamps and deuterium lamps are common sources of radiation in this region.

Some sort of wavelength selection is commonly provided in modern pho-

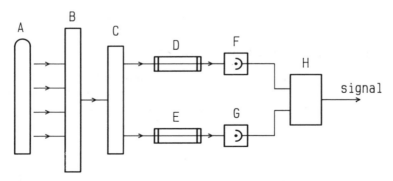

FIG. 8.1. Elements of photometric detection. A, Continuous light source; B, means of wavelength selection; C, beam splitter; D, detector cell containing sample; E, empty cell; F and G, photodetectors; H, signal processing.

tometers. This may be a simple system of filters that are interposed between the polychromatic source (a medium pressure mercury lamp) and the cell to provide wavelengths of 214, 220, 280, 313, 334, and 365 nm in addition to the intense 254-nm line.

A more versatile instrument is one that provides a continuous selection from the spectrum of radiation such as may be provided by a diffraction grating that may be rotated to give any desired wavelength. The ability to select any wavelength for examination is an extremely useful feature in that it allows the user to tailor a certain amount of specificity into a method. For example, if an analyte of interest has its maximum absorption at a wavelength where a nearby eluting band also has significant absorption then it will usually be profitable to monitor at a different wavelength where absorption of the interfering species is comparatively low.

The incident light is usually split into two beams, one of which irradiates the sample and the other an identical cell which is filled with either air or mobile phase. The transmitted light from this cell also impinges on a photodetector, and by appropriate signal processing the instrument delivers an output that represents the difference between the two photocells. This use of a comparator beam and detector is a very effective means for precisely nulling the high standing current from the transducer that examines the sample beam.

The detecting volume element is typically a small cylinder of effluent about 1 mm in diameter and 1 cm long. The volume of such a cell is on the order of 10 μl and represents a compromise between maximizing the path length to obtain greater response and avoiding excessive band broadening in the detector by keeping the volume as small as possible. In recent years as separation devices have become smaller it has become necessary to reduce the detecting volume still further so that nowadays it is possible to find a number of detectors where the cell volume is closer to 1 μl. The reader is directed to authoritative works on HPLC practice for further details on the construction and idiosyncrasies of spectrophotometer flow cells.[1]

8.2.2. Sensitivity of Spectrophotometers

Spectrophotometers are responsive only to those materials that absorb in the UV or visible range of the spectrum. There is in this the implication that such detectors can only be used to detect and therefore quantify samples that absorb in the UV or visible, and indeed some authorities explicitly state this to be the case. However, as we will see later, spectrophotometers can be very effectively employed to quantify eluites that are light-transparent although this still relies on their being responsive only to light-absorbing species.

The most common use of spectrophotometers and the context in which they display their greatest sensitivity is in the measurement of eluites that are light absorbers. To achieve the maximum in sensitivity the various components of the mobile phase should be transparent at the wavelength of measurement. When this condition prevails, the spectrophotometer becomes specifically sensitive to the eluites of interest.

How sensitive can the spectrophotometer become? The sensitivity of a detector used in this way is limited by its intrinsic noise. In a modern well-designed spectrophotometer this noise can be as low as 10^{-5} absorbance units so that by convention the limit of detection (LOD) will be that concentration of solute that generates a signal of 2×10^{-5} absorbance units. It is not uncommon for solutes to have molar absorptivities on the order of 10^5 liters mol^{-1} cm^{-1}, but using a value of 1000 as being more representative of the average and substituting in equation (8.2) where a cell length of 1 cm is assumed, one calculates the LOD of concentration to be 2×10^{-8} molar. By making some further assumptions about the width of a chromatographic peak (let us say 0.5 ml) and its shape (Gaussian or roughly triangular) one calculates the LOD of a solute with a molecular weight of 100 to be approximately 0.5 ng. For solutes of the highest molar absorptivities the limits of detection are impressive indeed.

The spectrophotometer used in this way as an eluite specific detector allows a great deal of flexibility in the choice of eluent composition. If the only constraint is that the mobile phase be essentially transparent then there is considerable latitude for choosing the composition of the mobile phase. For example, relatively concentrated electrolytes may be used in IC with photometric detection provided that they have very low absorption. This in turn permits the use of a wider range of exchanger capacities than may be possible with some other methods of detection. Changes in mobile phase pH and gradient elution using a wide range of eluting power are permissible as long as the mobile phase absorbance is kept very low.

All these features make photometric methods when used in this way a very versatile and powerful adjunct to IC when the sample species absorb in the UV or visible. This covers a great number of organic ions both cationic and anionic; ions with aromatic substituents or ions with conjugation usually have high molar absorptivities. Likewise many common inorganic ions have useful absorption properties at the shorter wavelength end of the UV spectrum as have a great number of complexed metal ion species.[2,3]

It is clear that situations may arise in IC where the choice lies between using a UV detector or one of the conductivity methods. There are any number of hypothetical conditions that one can imagine, but two come to mind where the preference for one detector over the other is easily argued. In the first case if a major peak poses a problem for conductometric detection of a closely eluting minor component then UV detection would be the logical choice if the minor component is UV absorbing and the major component is not. On the other hand, if a sample should contain nonionic materials that are strongly UV absorbing and tend to overlap the elution of ionic species of analytical interest then a conductometric method is indicated since the detector will be insensitive to the nonionic materials.

8.2.3. Photometric Detection with Postseparation Derivatization

If eluites lack chromophores and the mobile phase is itself transparent then photometers will obviously be insensitive to the elution of such species. It is possible, however, in some cases to generate derivatives of the ionic analytes that are chromophoric, by simple chemical reactions either before or after separation. The more common practice is after separation and there is a simple reason for this: there is greater freedom to change the composition of the mobile phase after separation when only the constraints of detection need be considered.

One of the earliest examples of this mode of detection is the classical work on the ion exchange separation of amino acids.[4] Here the effluent from the separator column(s) is mixed with ninhydrin reagent, which generates visibly colored complexes of the amino acids that are then examined by a photometer operating in the visible range of wavelengths. Transition metal ions separated by various IC methods are frequently measured after postseparation reaction with a variety of colorimetric reagents. Complexation reactions before separation can also be used to facilitate not only the separation but also the UV detection of certain metal ions. Denkert and co-workers[5] have coupled by ion pairing a UV-absorbing counterion with the transparent analyte ion to make the latter visible by way of its UV-detectable "flag" or companion ion.

The concept of reaction prior to detection is in principle a simple one and has been applied to a number of IC applications that will be described in Chapter 9.

8.2.4. Indirect Photometric Detection (IPD)

For some time it was widely held that UV–visible detectors would detect only those eluites that were either light absorbing or could be made to generate chromophores if they were not. In the late 1970s a number of workers showed how photometers could indeed be very effectively used to measure transparent eluites without modifying their chemistry in any way.[6-9] A major feature of this new approach to photometric detection was the use of light-absorbing (usually UV-absorbing) eluents, made so by including in the eluent light-absorbing ions of the same charge as the eluite ions. These light-absorbing ions have a dual role: of selectively displacing the analyte ions from the separator column and revealing them in the effluent. The appearance of eluites is signaled by the troughs that appear in the base-line absorbance as the transparent eluites substitute for the chromophoric displacing ions. Small and Miller,[7] who made an extensive study of the method, called it indirect photometric detection.

This advance in the use of photometers is made possible by the unique nature of the ion exchange reaction, which demands a strict linkage between the concentrations of eluite and displacing ions. This aspect of ion exchange has been treated elsewhere in the book but it is helpful to repeat its essential principles.

8.2.4a. Principles of Indirect Photometric Detection

Consider an ion exchange column—for illustrative purposes an anion exchanger—which has been pumped and equilibrated with salt solution denoted $Na^+ E^-$ so that the sites in the exchanger are occupied exclusively by eluent ions E^-. If an injection is made of sample electrolyte, denoted $Na^+ S^-$, then the eluite, S^-, will generally be retarded and appear at some characteristic elution volume determined by the usual factors. As S^- elutes there is a concerted drop in the concentration of E^- since electroneutrality and the equivalency of exchange demand that the total concentration of anions remains constant since the concentration of sodium co-ions is constant (see Figure 7.16). It follows therefore that the concentration of S^- in the effluent can be *indirectly* monitored by continuously monitoring the level of the eluent ion E^-. Thus, if sample ions are inconveniently lacking in a property, for example optical absorbance, one may exploit this deficiency by deliberately choosing an eluent ion that *is* light absorbing and monitoring the troughs generated in the base line by the transparent eluite ions. An example serves to illustrate how the method works.

A column containing an anion exchanger is equilibrated with a dilute (10^{-3} M) solution of sodium phthalate until the effluent absorbance is stable as indicated by a UV photometer monitoring the column effluent. A sample containing several anions yields a chromatogram such as depicted in Figure 8.2, where the troughs are due to the replacement of the ambient light-absorbing phthalate by the transparent eluite ions. The off-scale positive deflection is due to the ion exchange displacement of phthalate by the injected sample ions as a whole. In this example the total concentration of the sample exceeded that of the eluent so the void disturbance was positive. Had the total concentration of the sample been less than that of the eluent, the disturbance would have been negative.

8.2.4b. Sensitivity in IPD

In indirect photometric detection sample ions are revealed and quantified by the decrements they produce in eluent concentration. Since the displacing species is usually in much greater abundance than the eluite ions, these decrements will ordinarily represent rather small fractional changes in eluent level. Thus, the success of IPD is directly related to how precisely we can measure these fractional differences (the signal) in the presence of the random fluctuations (the noise) in the base-line response. Photometers used in this fashion therefore exemplify bulk property detectors and have the inherent limitations of this type of detector. Therefore it is critical to the understanding of IPD to appreciate how sensitivity is related to the concentration of the eluent.

Let us consider the case of elution of a sample ion through an anion exchanger. Typically this would result in a Gaussian-shaped change in sample ion concentration, but for simplicity of treatment we will assume that it appears as a

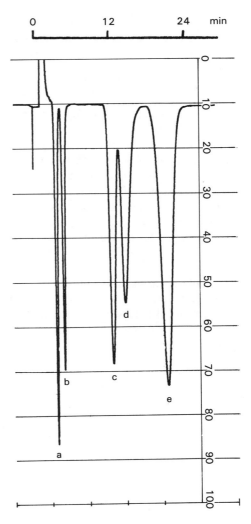

FIG. 8.2. Separation and indirect photometric detection of several "transparent" analyte ions: a, chloride; b, nitrite; c, bromide; d, nitrate; e, sulfate. Separator was a low-capacity anion exchange resin; displacing ion was phthalate. Ordinate is absorbance. (From Ref. 7 with permission. Copyright 1982 American Chemical Society.)

square wave pulse of concentration C_s. This pulse of eluite will cause a concomitant and identically shaped pulse change in the eluent ion level as indicated in Figure 8.3. The signal to be measured, S, is the difference between the base-line signal and the signal when the sample ion elutes.

This may be expressed as follows:

$$S = C_S \, \epsilon_S + (C_E - C_S) \, \epsilon_E - C_E \epsilon_E = C_S \, (\epsilon_S - \epsilon_E) \qquad (8.3)$$

where ϵ_S and ϵ_E denote the molar absorptivities of the sample and eluent ions, respectively, and C_E is the concentration of the eluent ion. Since the molar absorptivity of the sample species is presumed to be zero, equation (8.3) reduces to

$$S \propto C_S \epsilon_E \qquad (8.4)$$

This important relationship reveals that the response will be greater the greater the molar absorptivity of the eluent. But, as has been emphasized elsewhere in the book, response alone is not a sufficient measure of sensitivity; noise must be considered as well. The origins of noise in this type of measurement are not as readily identifiable as in, say, conductometric detection where the property, conductance, is so temperature sensitive. Absorbance is essentially temperature independent. Nevertheless there will be some source of base-line instability that will give rise to noise that we can assume to be proportional to the base-line signal level. This can then be conveyed by the simple expression

$$\text{Noise} \propto C_E \qquad (8.5)$$

It follows from (8.4) and (8.5) that the signal-to-noise ratio is proportional to the ratio $C_S \epsilon_E / C_E$, which means that sensitivity is enhanced by using more dilute eluents. The reader will recognize the strong parallel between this conclu-

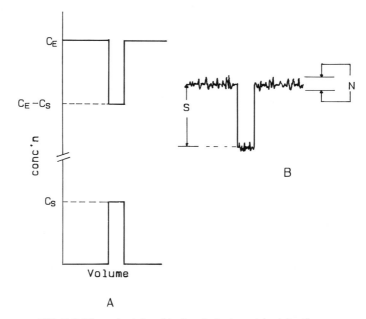

FIG. 8.3. The principle of indirect photometric detection.

sion and the one developed from a similar treatment of nonsuppressed conductometric detection. In both cases sensitivity is enhanced by using more dilute eluents. By the same token the IPD approach suffers from the same drawback as it pursues greater sensitivity: lowering the eluent strength increases the susceptibility to overload.

Since sensitivity is enhanced by lowering the concentration of E^- it has occasionally been proposed that sensitivity might be enhanced with no attendant loss in overload capacity by retaining E^- for its "eluite-revealing" role but placing the burden of the eluting role onto one or more other anions that are not chromophoric. There is a fallacy in this concept that is revealed in the following argument.

Assume that there are only two eluent anions denoted E^- and X^-, where E^- is light absorbing and X^- is not. For simplicity, let us further assume that their affinities for the anion separator are identical. If a certain concentration C of E^- is effective for the chromatographic separation in question then if we were to reduce the concentration of E^- tenfold we would have to provide X^- at a level of $0.9C$ to restore the elution power of the eluent to that of E^- alone. But would we have derived any benefit in increased sensitivity from the tenfold reduction in the concentration of E^-? It is true that the background absorbance would be reduced tenfold. On the other hand the decrement in eluent ion concentration is now shared between the two ions E^- and X^- and since they have the same affinities for the separator only one tenth of the impact of the eluite will be imposed on the concentration of E^-. Thus, even though the background noise will be one tenth of that using E^- alone, the response when using the mixture of E^- and X^- will be a tenth as much. So the net result is that the signal-to-noise ratio remains precisely the same, that is, no improvement in sensitivity.

A great merit of the IPD system is the degree of freedom that it allows in the choice of eluent ion. Virtually any ion that absorbs in the UV or visible is a possible candidate and it is usually some undesirable chromatographic feature that disqualifies it. For example, it might be too weak a displacing ion or equally undesirable it might be too potent in this respect. A wide variety of displacing ions have been used in the IPD of both cations and anions; a selection is given in Table 8.1.

IPD permits considerable flexibility in the choice of eluent concentration.

TABLE 8.1. Ions Useful as Eluents in Indirect Photometric Detection

For anions
  Nitrate, iodide, benzoate, phthalate, sulfobenzoate, trimesate, Cu(EDTA)[a]
For cations
  Cupric, benzyl trimethylammonium, copper(en)[a,b]

[a]Reference 10.
[b]en, ethylene diamine.

This is significant not just for improving sensitivity when analyte concentration is low but at the other end of the concentration scale where sample species may be abundant and sensitivity is of little concern. In these circumstances avoiding a dilution step will be of prime importance since it adds complexity to the analytical task as a whole. Then it will be advantageous to use ion exchangers of high capacity that can cope with the high ion content of a concentrated sample, in which case eluents of commensurately high ionic strength will be necessary. The potentially high absorbance of the eluent need pose no problem; by using a variable wavelength UV detector it will usually be possible to set the wavelength to a value where the molar absorptivity of the eluite ion is appropriately small thereby nullifying the concentration effect and providing a base-line absorbance that falls within the acceptable range of about 0.2–0.8 absorbance units. An example is provided in Chapter 9 that shows how IPD can handle a very concentrated sample quite successfully.

8.2.4c. Universal Calibration Feature of IPD

Calibration in IPD is accomplished in the usual way, by injecting known amounts of analyte ions and measuring either the depths or areas of the eluted troughs. It is common practice to reverse the polarity of the output signals so that chromatograms have a more conventional appearance with peaks rather than troughs. In harmony with this, IPD perturbations will be referred to as peaks, although from time to time the chromatograms may be presented in their natural downward-pointing direction if it is felt that it will facilitate the reader's understanding of the phenomena involved.

Small and Miller[7] showed that calibration plots of peak height versus analyte amount are linear over a usefully large range of concentration. Their work also revealed an interesting and useful aspect of calibration in the IPD mode, namely, that for many ions the area of the peak was independent of the ion injected and dependent only on the amount (cf. equation 8.4). This is a natural result of the method of monitoring since each equivalent of eluite displaces the same amount of monitoring ion from the mobile phase irrespective of the nature of the eluite ion. Small and Miller demonstrated this by making separate injections of accurate amounts of nitrate, sulfate, and phosphate into a sodium phthalate mobile phase at pH 8 using an anion exchange resin as stationary phase. The peak areas obtained are listed in Table 8.2 along with the peak area on an equivalent basis. The peak area was indeed approximately independent of the ion injected for the three ions measured and provides some basis for expecting that all ions would adhere to this rule. Anions of acids with medium to high pKs should give responses appropriate to their valence at the ambient pH of the eluent. Phosphate, for example, exists predominantly as the HPO_4^{2-} species at pH 8 so that one mole of phosphate injected would be expected to displace two

TABLE 8.2. Calibration Data for Nitrate, Sulfate, and Phosphate Determined by Indirect Photometric Detection[a]

Sample injected	Area of trough (arbitrary units)	Area of trough per meq of ion injected
5×10^{-3} M sodium nitrate	117.5	23.5
2.5×10^{-3} M sodium sulfate	111	22.2
1.67×10^{-3} M sodium o-phosphate	80.4	24.1

[a]Reference 7. Copyright 1982 American Chemical Society.

equivalents of monitoring ion. The data of Table 8.2 support this expectation. If eluite ions have appreciable absorbance on their own part at the monitoring wavelength then the rule of universal calibration will obviously be violated for that particular ion, and individual calibration is required, as is usually the case in chromatography.

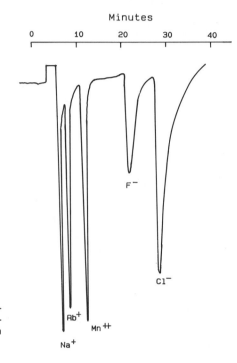

FIG. 8.4. Joint anion and cation determination by IPD. (From Ref. 7 with permission. Copyright 1982 American Chemical Society.)

8.2.4d. Simultaneous Measurement of Cations and Anions Using IPD

The work of Small and Miller demonstrated how IC in conjunction with IPD could provide a means of anion and cation determination using just one chromatographic run. They coupled an anion and cation exchange column together and, using an eluent, copper nitrate, that included both UV absorbing cations and anions, were able to obtain responses both when anions replaced the nitrate ions and cations the cupric ions from the eluent (Figure 8.4). As in other forms of chromatography this method requires minimum overlap of peaks for it to be most effective and this can involve considerable and rather complex "juggling" of eluent conditions to accomplish this. Iskandarani and Miller[11] developed a novel variant of IPD that permits coelution of anions and cations while providing separate anion and cation chromatograms, all from a single chromatographic run. They still used two columns, of cation and anion exchanger, to separate the appropriate target ions, and an eluent comprising cation and anion displacer ions that were both UV absorbing—copper sulfobenzoate in this case. The novelty of their approach was the use of a photometer that had the capability (1) to measure at two independently selectable wavelengths λ_1 and λ_2, two absorbances that will be denoted as A_1 and A_2; and (2) to provide a signal that was proportional to the quantity $A_1 - RA_2$, where R is a value that is selectable by the operator. The origin and significance of R will be apparent presently.

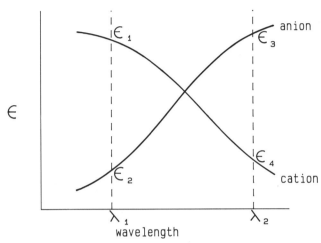

FIG. 8.5. Molar absorptivities (ϵ) as a function of wavelength for the anion and cation of a hypothetical IPD eluent, suitable for the simultaneous determination of cations and anions.

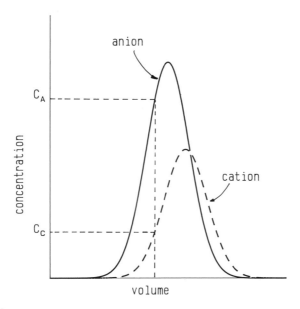

FIG. 8.6. Portion of a chromatogram showing the overlapping elution of cation and anion species.

To illustrate the principle of their method let us consider a hypothetical eluent where the molar absorptivities of its constituent anion and cation are represented by the plots of Figure 8.5. Let it further be assumed that the column effluent is monitored simultaneously at the two wavelengths λ_1 and λ_2, in which case the monitoring ions will have the molar absorptivities ϵ_1, ϵ_2, ϵ_3, and ϵ_4 as given by the intercepts of the vertical dashed lines with the two spectra.

Consider now a portion of a chromatogram (Figure 8.6) where there is significant overlap in the elution of an anion (solid line) and a cation. The signal at the volume represented by the vertical line will be the summation of two contributions: the decrements in absorbance due to the decrement in concentrations C_A and C_C caused by the presence of the anion and cation eluites.

We may express this signal in the following manner:

$$\text{Signal at wavelength } \lambda_1 = S_1 \propto C_A\epsilon_2 + C_C\epsilon_1 \tag{8.6}$$

and

$$\text{Signal at wavelength } \lambda_2 = S_2 \propto C_A\epsilon_3 + C_C\epsilon_4 \tag{8.7}$$

By rearranging and ratioing equations (8.6) and (8.7), we obtain the relationship

$$(S_1 - C_A\epsilon_2)/(S_2 - C_A\epsilon_3) = C_C\epsilon_1/C_C\epsilon_4 = \epsilon_1/\epsilon_4 \tag{8.8}$$

Now the quantity ϵ_1/ϵ_4 is readily determined from the spectra and we can assign it the term R_C so that equation (8.8) can now be written as

$$(S_1 - C_A\epsilon_2)/(S_2 - C_A\epsilon_3) = R_C \tag{8.9}$$

or, after rearrangement

$$S_1 - R_C S_2 = C_A(\epsilon_2 - R_C\epsilon_3) \tag{8.10}$$

Since the quantity $\epsilon_2 - R_C\epsilon_3$ on the right-hand side of the expression is a constant, then $S_1 - R_C S_2$ is proportional only to the concentration of anionic species at the appropriate elution volume. The photometer has by design the

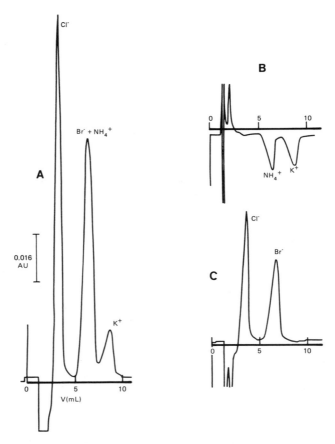

FIG. 8.7. Simultaneous determination of cations and anions. (A) Total chromatogram—note coelution of bromide and ammonium ions; (B) and (C) are, respectively, the cation and anion chromatograms obtained after signal processing. (From Ref. 11 with permission. Copyright 1985 American Chemical Society.)

capability of providing a signal that represents the quantity $S_1 - R_C S_2$ so it only remains to supply it with the value R_C and it will output a signal that is proportional *only to the amount of anion in the sample being examined*. In this fashion the photometer provides an anion chromatogram of the sample even though cations may be coeluting with some or all of the anions. By an identical line of reasoning to the above, it can be shown how the photometer will also deliver a chromatogram for the cations in the injected sample.

As well as being dependent on anion concentration, the magnitude of the signal $C_A(\epsilon_2 - R_C \epsilon_3)$ will obviously be determined by the magnitude of the constant $\epsilon_2 - R_C \epsilon_3$. There is no *a priori* means for maximizing this quantity for it depends on the spectra of the particular ion pair chosen. Therefore the selection of λ_1 and λ_2 requires some judgment on the part of the operator.

Figure 8.7 illustrates the application of this technique to a mixture of anions and cations where coelution of some of the species is present. The total chromatogram for Cl^-, Br^-, NH_4^+, and K^+ is shown in Figure 8.7A, where only three peaks are apparent. Monitoring at two wavelengths and operating on the signals in the way described above yields the separate chromatograms of Figures 8.7B and 8.7C, which clearly reveal the presence of four ionic species and how coelution of ammonium and bromide accounts for the three peaks of the total chromatogram.

IPD has also been used successfully in paired ion chromatopgraphy. It may also be applied to postseparation derivatization where the absorbance of the sensing species is not developed until after the separation step.

8.3. ELECTROCHEMICAL DETECTION

If species are electroreducible or electrooxidizable then it is possible to determine them with great sensitivity and selectivity by a suitable electrochemical device. The most common are amperometric detectors where a pair of electrodes bracket a small sample of solution, a potential is applied across the electrodes, and the current derived from the electrode reduction (oxidation) is a measure of the amount of electroactive species present. The potential may be applied as a constant dc potential or it may be pulsed. Pulsing can get around such problems as charging currents, and in one popular detector, the pulsed amperometric detector (PAD), voltage pulses of varying polarity and amplitude serve to cleanse the electrodes of the products of reaction and thereby stabilize their characteristics.[12,13]

The ability to control applied potential affords a measure of specificity in that it is possible to desensitize the amperometric detector to many species when their redox characteristics are known. The requirement that the solution between the electrodes be conducting is easily met in ion chromatography.

Because the electrochemical detectors rely on surface chemical reaction—

TABLE 8.3. Some Electroactive Species That Have Been Determined by Ion Chromatography

---

HS^-, CN^-, Br^-, I^-, SCN^-, SO_3^{2-}, $S_2O_3^{2-}$, OCl^-, hydrazine, alcohols, glycols, phenols, catechols, monosaccharides, disaccharides, oligosaccharides

---

the reaction at an electrode—their behavior is often more complex and they are more subject to interferences than detectors that do not involve an interface as a critical part of the detection process. Thus the construction of the electrodes, their composition, their susceptibility to poisoning and subtle participation of species that are not electroactive are just some of the key factors in electrochemical detection. Given the breadth of the subject and the scope of this text it must suffice to refer the reader to texts and articles on electroanalytical chemistry for appropriate background reading.[14,15] Users of electrochemical detectors can also expect to gain useful guidance from the product literature of the manufacturers.

Table 8.3 lists some of the species that have been determined by electrochemical detectors. For some ions such as cyanide and sulfide it can be the only effective method. Suppressed IC is insensitive to these anions of very weak acids, and while nonsuppressed IC or indirect photometric detection may be used, the specificity of electrochemical detection may be ideally suited to dealing with closely eluting ions that would interfere in these other methods.

8.4. MISCELLANEOUS METHODS

Various other techniques have been examined as alternatives to the major methods of detection in IC. Dasgupta[16] has provided an excellent summary of the various techniques that have been tried or used up to about 1984. Most of the common detectors have been tried in the indirect mode: refractometer, [17] fluorometer,[18–20] and electrochemical.[21] Haddad and Heckenberg[17] claim superior sensitivity for indirect refractometric detection compared to IPD, while Jenke and co-workers[22] claim the opposite to be the case.

Downey and Hieftje[23] described a technique that they called replacement ion chromatography (RIC). A quotation from their paper summarizes the essential principles of their approach: "Replacement ion chromatography is conceptually an approach whereby sample ions are first separated chromatographically, but are then stoichiometrically replaced in the eluent by an ion that can be detected more sensitively than the sample ion itself." In the examples used by these workers "lithium ions are used to replace the counterions associated with sample anions. . . . Alternatively, sample cations can be directly replaced with lithium ions. . . . Once replacement is complete, the eluent is directed to a flame photometer dedicated to lithium measurement. Chromatograms are therefore simply a plot of lithium concentration as a function of time.''

The source of lithium ions is a small cation exchange bed in the lithium form placed after the separator column. Of course eluent ions as well as sample ions would replace lithium to give high background signal in the flame photometer so a suppressor column must be used to remove the eluent species and allow the analytes to pass in exactly similar fashion to suppressed conductometric analysis. Downey and Hieftje cited as major advantages of the RIC approach using flame photometry the very low detection limits for lithium, and the facts that only a single ion species (lithium) is needed for calibration and the detector is relatively insensitive to temperature.

In practice RIC like other methods reveals its limitations. For example, the carbonic acid from the suppressor displaces some lithium from the ion replacement bed and a significant background signal is therefore unavoidable. Using sodium hydroxide eluents alleviates the background ''bleed'' of lithium from the replacement bed but does not entirely eliminate it. Cation determinations are of course also possible by this approach. Impurities in the eluents can be a cause of lithium bleed but even if one could eliminate them entirely, the ionization of water, albeit that it is relatively minute, nonetheless defines the ''ceiling'' of detectability. In comparing RIC to suppressed IC the authors claim comparable or superior detection limits for the former method but no systematic comparison has been reported.

Postcolumn reaction with silver nitrate reagent followed by turbidimetric detection using a simple UV spectrophotometer provided an anion exchange method with specificity for halide ions as well as species such as thiocyanate and chromate ions.[24]

Radiometric methods of detection have been employed for many years in ion exchange studies as well as in ion exchange chromatography. Radiolabeling of ions is an extremely effective means of following their distribution between phases and has been extensively used to study both the thermodynamics and kinetics of ion exchange phenomena. In chromatography, radiolabeling may be used to confirm or to ascertain the elution behavior of a species, but on the whole the applicability of radiometric methods to nonradioactive species is limited as compared to other methods of detection. However, if the eluites are radioactive then radiometric detection techniques are not only very powerful but in fact are most likely to be the only method with the sensitivity adequate to the task. Since the practice of radiometric detection is quite specialized, the reader is referred to the original literature for details of applications. In this respect, the publications of Horwitz and co-workers[25,26] on the ion exchange separation of transuranic elements is recommended.

8.5. DETECTION IN IC—A SUMMARY

A great many detection schemes have been proposed or tried for ion chromatography and several are in widespread use. Inevitably there has also been a proliferation of names applied to these techniques: suppressed, nonsuppressed, direct, indirect, solute specific, bulk property, and so on. The names are not always apt nor is there general agreement on what some of them mean. In view of this confusion, a treatment that stresses principles is preferable to one that relies on names and rigid definitions. The aim of this summary is to identify the unifying principles of detection in IC and to link them to names where appropriate. Since ion exchange is the principal separating mode in IC, an ion exchange elution will be assumed in the illustration.

When an ion exchange column has been equilibrated with mobile phase the equivalent concentration of displacing or eluent ion(s) in the effluent is constant and is denoted as C in Figure 8.8. (For simplicity, a single displacing species is assumed.) When eluite appears in the column effluent, there is a sympathetic drop in the concentration of the eluent ion. The requirements of ion exchange equivalence and electroneutrality dictate that the sum of the concentrations of eluite and eluent ions shall remain at all times equal to C.

Let us assume that a detector placed downstream from the column is set to

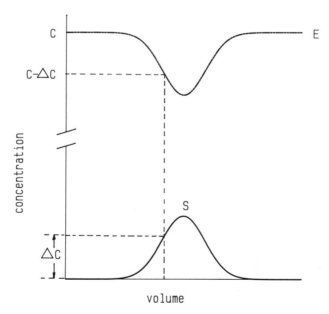

FIG. 8.8. A fundamental feature of detection in ion exchange chromatography. Analyte species S causes a concerted drop in the concentration of displacing species E.

monitor a property denoted P and further that this property is possessed to some extent by both the eluent ion and the eluite from an injected sample. P_E and P_S denote the values of the properties per equivalent of the respective ionic species.

The ambient or background signal from the effluent in the absence of sample species can therefore be represented as

$$CP_E \tag{8.11}$$

since it is assumed that the expressed signal is proportional to the property.

The signal at any time during the elution of a sample ion is in turn represented by

$$(C - \Delta C)P_E + \Delta CP_S \tag{8.12}$$

The detector response at a particular instant during the elution of an eluite is the difference between the background signal and the signal at that instant which is proportional to the difference between (8.11) and (8.12), that is

$$\text{Response} \propto \Delta C(P_S - P_E) \tag{8.13}$$

A number of conclusions can be developed from this simple expression:

1. The detector will respond to the eluite if and only if $P_S \neq P_E$.
2. Response can be positive with respect to the background ($P_S > P_E$) or negative ($P_E > P_S$).

Case I ($P_S >>> P_E$). Many important detection methods employ eluents that are essentially lacking in the property possessed by the analytes; P_E tends to zero in other words. UV, electrochemical, radiochemical, and fluorescence detectors used in the conventional way are good examples. Suppressed conductometric detectors reduce the conductivity of the effluent before passing it to the transducer and in effect also belong to this class. These are sometimes referred to as direct detection methods and the detectors as solute-specific detectors.

Case II ($P_S <<< P_E$). Response can be elicited from analytes when eluent ions are rich in the property relative to the analytes. In these cases responses are negative relative to the background. Indirect photometric detection using light-absorbing eluent ions to displace transparent analyte ions is a good example of this. Theoretically, any property can be exploited in this way and many have been. These methods are frequently referred to as indirect methods.

Case III ($P_S > P_E$ or $P_S < P_E$). Responses can be positive or negative within the same method. Nonsuppressed conductometric detection is the best known example of this. When hydronium ions or hydroxide ions are employed to displace cations or anions, respectively, the responses are negative since the equivalent conductances of these ions are greater than those of any analyte ions. With much less conducting ions such as phthalate or benzoate, the responses can be positive or negative depending on the relative magnitudes of the equivalent conductances of analyte and eluent.

It is possible to obtain both positive and negative analyte responses in the same chromatogram.

3. Response will be independent of the nature of the sample ion and dependent only on its concentration when $P_S = 0$. Indirect photometric detection is a good example of this. In most cases in IPD the sample ions are nonabsorbing while the eluent ion is absorbing and the injection of equivalent amounts of transparent analytes gives peaks of the same area.

To be measurable, an analyte must elicit a response in the detector. Sensitivity of detection depends on how the response compares to the average noise of the background signal. Noise is roughly proportional to the background signal, so methods that provide for a low background signal in the detector tend to be more sensitive.

REFERENCES

1. L. R. Snyder and J. J. Kirkland, *Introduction to Modern Liquid Chromatography*, 2nd edition, Wiley-Interscience, New York (1979.
2. R. P. Buck, S. Singhadeja, and L. B. Rogers, Ultraviolet Absorption Spectra of Some Inorganic Ions in Aqueous Solutions, *Anal. Chem.* **26**, 1240–1242 (1954).
3. R. J. Williams, Determination of Inorganic Anions by Ion Chromatography with Ultraviolet Absorbance Detection, *Anal. Chem.* **55**, 851–854 (1983).
4. D. H. Spackman, W. H. Stein, and S. Moore, Automatic Recording Apparatus for Use in the Chromatography of Amino Acids, *Anal. Chem.* **30**, 1190–1206 (1958).
5. M. Denkert, L. Hackzell, G. Schill, and E. Sjogren, Reversed-Phase Ion-Pair Chromatography with UV-Absorbing Ions in the Mobile Phase, *J. Chromatogr.* **218**, 31–43 (1981).
6. A. Laurent and R. Bourdon, Dosage des anions par chromatographie d'échange d'ions, *Ann. Pharm. Fr.* **36**(9–10), 453–460 (1978).
7. H. Small and T. E. Miller Jr., Indirect Photometric Chromatography, *Anal. Chem.* **54**, 462–469 (1982).
8. R. A. Cochrane and D. E. Hillman, Analysis of Anions by Ion Chromatography Using Ultraviolet Detection, *J. Chromatogr.* **241**, 392–394 (1982).
9. D. R. Jenke, Modeling of Analyte Behavior in Indirect Photometric Chromatography, *Anal. Chem.* **56**, 2674–2681 (1984).
10. K. Hayakawa, T. Sawada, K. Shimbo, and M. Miyazaki, Effect of (Ethylenediaminetetraaceta-to)copper(II) and Bis(ethylenediamine)copper(II) Eluents on Nonsuppressed Ion Chromatography with Indirect Photometric Detection, *Anal. Chem.* **59**, 2241–2245 (1987).
11. Z. Iskandarani and T. E. Miller Jr., Simultaneous Independent Analysis of Anions and Cations Using Indirect Photometric Chromatography, *Anal. Chem.* **57**, 1591–1594 (1985).
12. S. Hughes, P. L. Meschi, and D. C. Johnson, Amperometric Detection of Simple Alcohols in Aqueous Solutions by Application of a Triple-Pulse Potential Waveform at Platinum Electrodes, *Anal. Chim. Acta.* **132**, 1–10 (1981).
13. S. Hughes and D. C. Johnson, Amperometric Detection of Simple Carbohydrates at Platinum Electrodes in Alkaline Solution by Application of a Triple-Pulse Potential Waveform, *Anal. Chim. Acta* **132**, 11–12 (1981).

14. P. T. Kissinger, Amperometric and Coulometric Detectors for High Performance Liquid Chromatography, *Anal. Chem.* **49,** 447A (1977).
15. P. T. Kissinger, Electrochemical Detection in Liquid Chromatography and Flow Injection Analysis, in *Laboratory Techniques in Electro-analytical Chemistry* (P. T. Kissinger and W. R. Heineman, eds.), Marcel Dekker, New York (1984).
16. P. K. Dasgupta, Approaches to Ionic Chromatography, in *Ion Chromatography* (J. C. Tarter, ed.), Marcel Dekker, New York (1987).
17. P. R. Haddad and A. L. Heckenberg, High Performance Liquid Chromatography of Inorganic and Organic Ions Using Low-Capacity Ion-Exchange Columns with Indirect Refractive Index Detection, *J. Chromatogr.* **252,** 177–184 (1982).
18. Sun-Il Mho and E. S. Yeung, Detection Method for Ion Chromatography Based on Double-Beam Laser-Excited Indirect Fluorometry, *Anal. Chem.* **57,** 2253–2256 (1985).
19. K. Jurkiewicz and P. K. Dasgupta, Positive-Signal Indirect Fluorometric Detection in Ion Chromatography, *Anal. Chem.* **59,** 1362–1364 (1987).
20. J. H. Sherman and N. D. Danielson, Indirect Cationic Chromatography with Fluorometric Detection, *Anal. Chem.* **59,** 1483–1485 (1987).
21. J. Ye, R. P. Baldwin, and K. Ravichandran, Indirect Electrochemical Detection in Liquid Chromatography, *Anal. Chem.* **58,** 2337–2340 (1986).
22. D. R. Jenke, P. K. Mitchell, and G. K. Pagenkopf, Use of Multiple Detectors and Stepwise Elution in Ion Chromatography without Suppressor Columns, *Anal. Chim. Acta.* **155,** 279–285 (1983).
23. S. W. Downey and G. M. Hieftje, Replacement Ion Chromatography with Flame Photometric Detection, *Anal. Chim. Acta* **153,** 1–13 (1983).
24. H. Small, unpublished results.
25. E. P. Horwitz and C. A. A. Bloomquist, The Preparation, Performance and Factors Affecting Band Spreading of High Efficiency Extraction Chromatographic Columns for Actinide Separations, *J. Inorg. Nucl. Chem.* **34,** 3851–3857 (1972).
26. E. P. Horwitz and C. A. A. Bloomquist, High-Speed Radiochemical Separations by Liquid–Liquid Chromatography Using a CSP Support. 1. Performance of Quaternary Ammonium Chloride–Zipax System for the Elution of CdII with HCl, *J. Chromatogr. Sci.* **12,** 11–22 (1974).

Chapter 9

Selected Applications of Ion Chromatography

9.1. INTRODUCTION

Previous chapters have been concerned with two of the major tasks of ion chromatography: how to effect an adequate separation of ions in a sample and how to detect them in the column effluent. This chapter provides some examples from the practical art of ion chromatography that illustrate how these two problems have been solved in a mutually compatible way. The examples are grouped into a number of topics, and, while the choice of topics and examples is somewhat arbitrary, there has nevertheless been a deliberate effort to pick cases that may best illustrate some of the basic principles of IC discussed earlier in the book. The chapter is not intended to be a review of the voluminous literature on the applications of IC.

9.2. CONDUCTOMETRIC DETECTION IN THE IC OF COMMON INORGANIC ANIONS

Conductometric detection is one of the most widely used procedures in IC. It is routinely employed in the determination of a great variety of ionic species, organic as well as inorganic, although initially it was targeted at inorganic analysis and more specifically at inorganic anions for which then existing methods were slow and often insensitive. Ion chromatography with conductometric detection radically changed the analytical chemistry of such common anions as fluoride, chloride, bromide, nitrate, nitrite, sulfate, and phosphate and a host of other less common species. Several reviews[1–4] discuss these many applications of IC; this section will illustrate with a few examples the present-day capabilities of conductometric methods.

Early applications are worth recording for they provide unique benchmarks

213

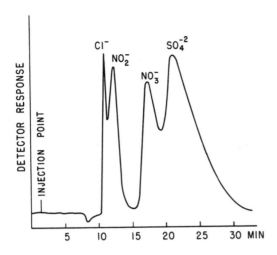

FIG. 9.1. Early separation of anions using sodium hydroxide as eluent and a column suppressor.

against which to measure progress. Figure 9.1 represents early attempts to separate a larger group of anions using sodium hydroxide as eluant. Peak shape of eluted ions was generally poor and some high affinity ions such as phosphate and iodide had such excessively long elution times that hydroxide eluents were impractical. This was a stumbling block to the development of anion IC until the demonstration in late 1972 that eluents containing the phenate ion were effective in displacing a great number of inorganic and organic ions of widely varying selectivities (Table 7.2). Phenate-containing eluents were used for anion analysis until the discovery of the carbonate eluent system in late 1974. Figure 9.2 represents the status of anion IC in 1975 using carbonate eluents and a packed bed suppressor.[5]

Since 1975 there have been many and marked improvements in the capabilities of suppressed conductometric IC. As a result of improved resolving power, separations are now considerably faster, sensitivity is much better, and a greatly expanded spectrum of ion affinities can now be handled in a single chromatographic run. Most of these advances have hinged on two key developments: improved resins and membrane suppressors.

The improvement in resins and column efficiency is exemplified by the chromatogram of Figure 9.3, which shows the separation of seven inorganic ions in less than 2 min. Performance such as this not only leads to faster analysis but the higher efficiency also means improved sensitivity.

A criticism of suppressed IC in its early days was that suppressors added dead volume that degraded efficiency. The system used for this separation employed the latest type of low dead volume suppressor, the MicroMembrane Suppressor. The quality of the resolution demonstrates the efficiency that is now achievable using modern suppressor technology.

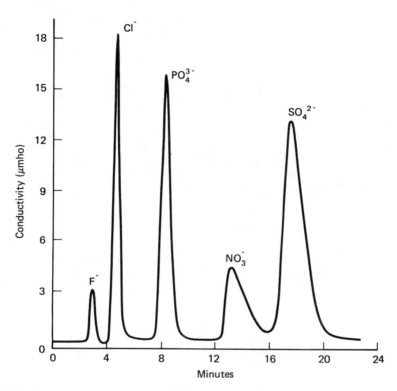

FIG. 9.2. Separation of anions *circa* 1975. Carbonate–bicarbonate eluent; suppressed conductometric detection—column suppressor. (From Ref. 5 with permission.)

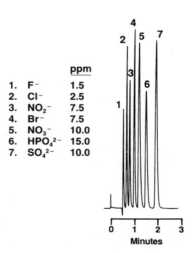

		ppm
1.	F^-	1.5
2.	Cl^-	2.5
3.	NO_2^-	7.5
4.	Br^-	7.5
5.	NO_3^-	10.0
6.	HPO_4^{2-}	15.0
7.	SO_4^{2-}	10.0

FIG. 9.3. Separations of anions *circa* 1988. Eluent: 2.0 mM Na$_2$CO$_3$, 0.15 mM NaHCO$_3$. Detection: suppressed conductometric; Anion Micro-Membrane (AMMS). (From Ref. 6 with permission.)

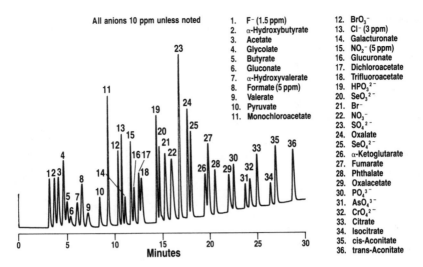

All anions 10 ppm unless noted

1. F⁻ (1.5 ppm)
2. α-Hydroxybutyrate
3. Acetate
4. Glycolate
5. Butyrate
6. Gluconate
7. α-Hydroxyvalerate
8. Formate (5 ppm)
9. Valerate
10. Pyruvate
11. Monochloroacetate
12. BrO_3^-
13. Cl⁻ (3 ppm)
14. Galacturonate
15. NO_2^- (5 ppm)
16. Glucuronate
17. Dichloroacetate
18. Trifluoroacetate
19. HPO_3^{2-}
20. SeO_3^{2-}
21. Br⁻
22. NO_3^-
23. SO_4^{2-}
24. Oxalate
25. SeO_4^{2-}
26. α-Ketoglutarate
27. Fumarate
28. Phthalate
29. Oxalacetate
30. PO_4^{3-}
31. AsO_4^{3-}
32. CrO_4^{2-}
33. Citrate
34. Isocitrate
35. cis-Aconitate
36. trans-Aconitate

FIG. 9.4. Gradient elution of inorganic and organic anions on a pellicular anion exchange resin. Eluent: gradient of 0.75 mM to 100 mM NaOH. Detection: suppressed conductometric; Anion MicroMembrane Suppressor. (From Ref. 7 with permission.)

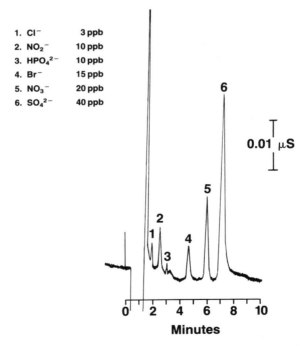

1. Cl⁻   3 ppb
2. NO_2^- 10 ppb
3. HPO_4^{2-} 10 ppb
4. Br⁻   15 ppb
5. NO_3^- 20 ppb
6. SO_4^{2-} 40 ppb

0.01 μS

FIG. 9.5. Determination of trace amounts of anions by direct injection. Eluent: 2.8 mM NaHCO₃, 2.2 mM Na₂CO₃. Detection: suppressed conductometric; Anion Micro-Membrane Suppressor. Sample size: 100 μl. (Reproduced courtesy of Dionex Corporation.)

The separation of Figure 9.4 is an impressive example of the capabilities of present day suppressed conductometric detection. The eluent is sodium hydroxide, applied as a gradient and suppressed in a MicroMembrane device. The separation is remarkable for the wide range of ions being measured, organic as well as inorganic, and for the fact that this is accomplished with hydroxide as the displacing ion. That this is possible is due in great part to the dynamic capacity of the membrane suppressor to cope with the relatively large concentrations of eluent required to displace the more tightly held ions. Theoretically, gradient elutions of this type are possible using packed bed suppressors, but the penalty would be a severely abbreviated lifetime before regeneration is necessary.

IC with suppressed conductometric detection is capable of great sensitivity. This is conveyed to some extent by Figure 9.5, which shows a chromatogram obtained from the direct injection of a 100-μl sample; much lower limits of detection are possible by using concentrator columns to collect the contained anions from large volumes of extremely dilute sample.

Figures 9.6 and 9.7 show examples of conductivity detection being used without suppression. In this case the effluent from the separator passes directly to the conductivity cell. Anions are typically separated on low-capacity anion exchange resins using a variety of eluents. Sodium hydroxide and benzoate are suitable for displacing anions of low affinity such as chloride, acetate, and

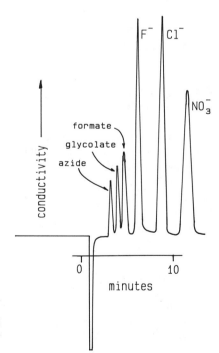

FIG. 9.6. Separation of anions on a low-capacity anion exchange resin. Eluent: 0.001 *M* benzoic acid. Detection: nonsuppressed conductometric. (From Ref. 8 with permission.)

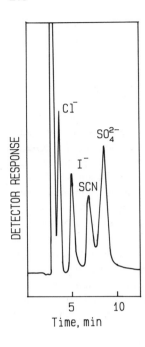

FIG. 9.7. Separation of anions on a low-capacity anion exchange resin. Eluent: 10^{-4} M potassium biphthalate; nonsuppressed conductometric detection. (From Ref. 9 with permission.)

nitrite, but for more tightly held species such as sulfate, iodide, and thiocyanate, an eluent such as sodium phthalate provides the requisite displacing power.

9.3. ANALYSIS OF ALKALI METAL AND ALKALINE EARTH METAL ION MIXTURES

The determination of alkali metal ions, particularly sodium and potassium, and of the alkaline earth ions calcium and magnesium is a common problem. It crops up in the analysis of brines, surface waters, soil, plant extracts, foodstuffs, etc. Ion chromatography is now a widely accepted analytical procedure in all of these areas. We will examine some of the chromatographic methods that have been developed for they provide useful illustration of how the basic principles of separation and detection are applied to the solution of a challenging chromatographic problem.

9.3.1. Alkali Metal and Alkaline Earth Metal Ions Together

The chromatographic analysis of alkali metal–alkaline earth ion mixtures is relatively straightforward when only one of the groups of ions is to be determined; it is more complicated when it is necessary to determine both groups in the same sample with but a single chromatographic run. The problem is the wide difference in selectivity between monovalent and divalent cations for the ion exchange resins normally used to separate them. Some simulated chromatographic separations can help define the problem.

These experiments assume the use of a 2-ml column of low-capacity cation exchanger (0.01 meq/ml) and hydrochloric acid as eluent. It will be assumed that the column has an efficiency of 3000 plates. We will use the equilibrium selectivity constants of Table 4.1 for the various hydronium ion–metal ion exchanges. Elution volumes of the various ions are calculated by substitution of the appropriate values in relationships (4.22) and (4.28) and (2.10) and assuming that the capacity of a fully sulfonated resin (C_R) is 2 meq/ml.

The first simulation (Figure 9.8A) uses a concentration of acid (0.02 M) that gives a complete separation of all four of the ions. This relatively low acid concentration is necessary to adequately resolve sodium and potassium but at the same time it leads to a prolonged elution of the divalent cations. From a practical standpoint this has two undesirable features: first, and obviously, the analysis time is undesirably long—greater than 50 min per sample if a flowrate of 2 ml/min is assumed; second, the sensitivity for the divalent ions is degraded because of the extensive band spreading. Analysis time can be considerably

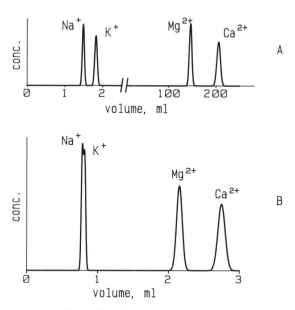

FIG. 9.8. Simulated separation of sodium, potassium, magnesium, and calcium on a low-capacity cation exchanger (0.01 meq/ml). (A) Using 0.02 M HCl as eluent. (B) Using 0.2 M HCl as eluent.

shortened by increasing the acid concentration in the mobile phase (Figure 9.8B) but now resolution of sodium and potassium is lost.

One solution to the problem would use a dilute acid in the early stages of the elution to resolve sodium and potassium and make a step change to the more concentrated acid to resolve magnesium and calcium and elute them within an acceptably short time. However, in developing a total method it is also necessary to consider what implication a particular separation strategy has for detection.

Since conductometric detection is a logical choice for these non-chromophoric ions the first step is an acceptable one; the dilute acid eluent does not put an undue load on any suppressor that might be used in the suppressed conductance mode and it favors sensitivity in nonsuppressed detection. Switching to a more concentrated acid eluent is not, however, an acceptable solution for it can overburden a suppressor and it will certainly greatly diminish the sensitivity for detection if nonsuppressed detection is employed.

An acceptable strategy, and one that is widely used, is to switch to a dilute solution of a more potent displacing ion after enough dilute acid eluent has been applied to resolve the monovalent ions. In suppressed detection, an eluent

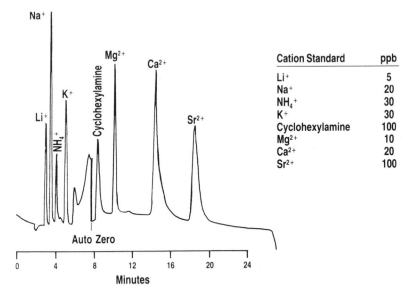

FIG. 9.9. Separation of mono- and divalent cations on a low-capacity cation exchanger using two eluents. Eluent 1: 12 mM HCl, 0.5 mM DAP·HCl. Eluent 2: 48 mM HCl, 8.0 mM DAP·HCl. Detection: suppressed conductometric; Cation MicroMembrane Suppressor. (From Ref. 10 with permission.)

that is currently used is a dilute solution of di-amino propionic acid (DAP). In a hydrochloric acid environment this amphoteric species exists as a doubly positive cation and as such is very effective in displacing divalent cations. In the alkaline environment of the suppressor the acid is converted to its anion, $NH_2CH_2CH(NH_2)COO^-$, which is then effectively removed from the effluent by the cation suppressor membrane–regenerant system. Figure 9.9 illustrates this type of system being applied to the determination of trace levels of alkali metals and alkaline earth ions. The elution in this case was preceded by a concentration step in which 5 ml of sample were first loaded from water onto a small concentrator column, which was also a low-capacity cation exchanger. The levels of lithium, sodium, ammonium, potassium, magnesium, calcium, and strontium in the 5-ml sample were 5, 20, 30, 30, 10, 20, and 100 parts per billion, respectively. As well as illustrating a common method for this type of analytical problem, this example also illustrates the use of ion exchange concentration as a means of enhancing detectability limits in IC.

Nonsuppressed detection also uses the "more potent eluent ion" strategy for eluting the divalent ions. In this case the diprotonated ethylene diamine species provides the requisite displacing power at the low concentrations necessary for adequate sensitivity.

9.3.2. A "Mechanical" Solution

A recent method for the simultaneous determination of the two groups of ions uses only a single eluent but employs what might be called a "mechanical" device to overcome the problem of their widely different elutability. The system is illustrated schematically in Figure 9.10. It uses two columns in series that are separated by a three-way valve. The exit of the second column is also connected to a three-way valve, which in turn is connected to a suppressed conductance detector. The remaining two inlets of the valves are connected together. Two valve configurations are used in a chromatographic run. In the first, the eluent flow is directed through both columns to the detector, while in the second, eluent flow is directed through the first column but the effluent from that column then bypasses the second column and passes directly to the detector.

Both columns contain low-capacity cation exchange resins. Their capacity and dimensions are designed so that their combined capacity is enough to resolve all of the monovalent species, while the first column alone contains enough capacity to resolve the divalent ions. The column capacities are adjusted in such a way that when the last eluting monovalent ion has just cleared the detector, the lowest affinity divalent ion, magnesium, is about to elute from the first column. At this point in the run, the valves are then switched to the second configuration thus directing the effluent from the first column directly to the detector where the resolved divalent ions may be measured. Diverting the eluent stream in this way

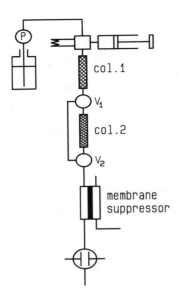

FIG. 9.10. Arrangement for the split column method for the determination of alkali and alkaline earth metal ions. Columns 1 and 2 are cation exchange resins. Valves V_1 and V_2 are three-way valves that allow column 2 to be bypassed. In this schematic, suppressed conductometric detection is assumed, but other methods of detection may be used instead.

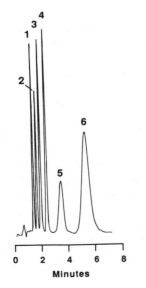

1. Li^+
2. Na^+
3. NH_4^+
4. K^+
5. Mg^{2+}
6. Ca^{2+}

FIG. 9.11. Split column determination of cations. Eluent: 20 mM HCl, 0.3 mM DAP·HCl was pumped through two columns (see Figure 9.10) for about 3 min to elute the alkali metal ions and the ammonium ion. Then the second column was bypassed and the effluent from column 1 was passed directly to the detector. Detection: suppressed conductometric. (Courtesy Dionex Corporation.)

avoids the unnecessary retention and bandspreading of the divalent ion bands in the second part of the column train. Figure 9.11 is an example of the split column approach.

While mechanical solutions such as this may not have the elegance of a chemical solution to a problem, they are nonetheless very effective. With modern instruments that use automated valves and where the timing of valve switching can be preprogrammed and controlled by a computer, mechanical solutions are relatively easy to implement and when "tuned up" require little intervention by the operator.

9.3.3. Alkali Metal or Alkaline Earth Metal Ions Separately

When both groups of ions are present in the same sample but the analytical interest is in only one of the groups, the problems are simpler. If a determination of only the divalent ions is required then the solution is very straightforward— elution with a DAP eluent, for example, will give a fast elution of the alkaline earth ions. The monovalent ions will elute early and unresolved but that does not matter since, in the problem as defined, they are of no analytical interest.

The situation is slightly more complicated if we wish to determine the alkali metals but not the alkaline earth species. A dilute acid eluent will resolve the monovalent cations but when the essential part of the chromatography is complete the separator still contains any divalent ions that were injected with the sample. This may not pose any problem for the next sample injected, but, with repeated sample injections, the divalent ions from the first and subsequent samples will eventually coelute with the monovalent ions from later samples. This can obviously cause significant errors in the determination of the monovalent ions. Although it did not use conductometric detection, the example of Figure 9.12 shows one solution to this problem.

It was necessary to determine low concentrations of sodium and potassium in 3% calcium chloride solution. Instead of a low-capacity resin a small column of fully sulfonated resin was used to take advantage of its high overload capacity since undiluted samples were to be injected. The mobile phase was dilute copper sulfate, and indirect photometric detection, using a UV photometer set at 252 nm, revealed the sodium and potassium as they displaced the UV-absorbing cupric ion in the effluent. In addition to the usual sample injection valve, the train contained another injection valve equipped with a much larger capacity loop (about 1 ml). This loop was charged with a concentrated solution of copper nitrate. When the chromatographic separation was complete the calcium that remained on the column was rapidly removed by injecting the 1-ml "shot" of copper nitrate solution. This of course caused a momentary large disturbance on the detector but that subsided rapidly and the column and detector returned to their original conditions ready for the injection of another sample.

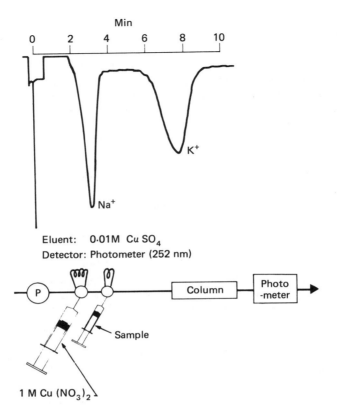

FIG. 9.12. Determination of sodium and potassium in a calcium chloride brine.[11] Example of "purge" elution of tightly bound divalent ions after ions of analytical interest have eluted. Detection: indirect photometric.

9.4. IC OF HIGH-AFFINITY IONS

The term "high-affinity ions" is used in this context to describe analyte ions that are difficult to elute with the common eluent ions. This high affinity can be due to the high charge of the eluites, or to other strong interactions that force them onto the resin phase. They can be problematical for some IC methods and require special solutions. Here are some examples.

9.4.1. Polyphosphates and Similar Species

Polyphosphates are widely used, notably in detergent formulations, so there is a need for methods to determine them over a wide range of concentrations.

They posed a problem for early conductometric IC because of the very high charge they acquired in the typical basic eluent. Quite concentrated eluents were necessary to dislodge them from the anion exchange resin separator columns which shortened the lifetime of packed bed suppressors or overloaded the early fiber suppressors. Chromatographers found an alternative method of detection by taking advantage of the spectral properties of ferric complexes of polyphosphate

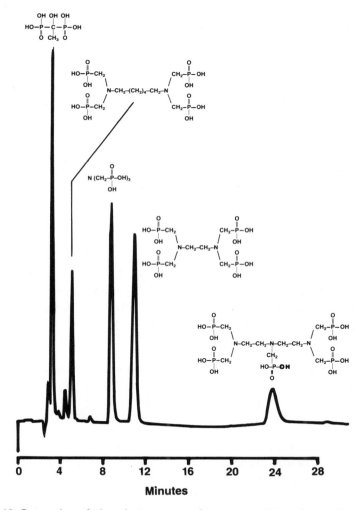

FIG. 9.13. Separation of phosphate sequestering agents on an anion exchange resin. Detection: postcolumn reaction with ferric ion followed by photometric detection of colored complexes. (From Ref. 12 with permission.)

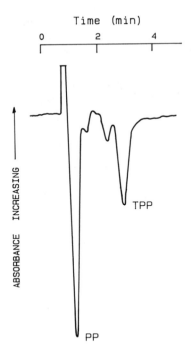

FIG. 9.14. Separation of pyro- and tripoly-phosphate ions on a pellicular anion exchange resin using trimesate ion as the displacing ion and indirect photometric detection (from Ref. 13 with permission. Copyright 1982 American Chemical Society.)

ions. They separated the polyphosphates on an anion exchanger but then mixed the effluent with a solution of a ferric salt and passed it to a UV–visible detector which measured the colored complex. Figure 9.13 illustrates the determination of some highly charged phosphate species by this approach. The stationary phase was a typical pellicular anion exchanger and the eluent a dilute nitric acid solution. Employing an acid as eluent subdued the charge of the polyanions, which in turn promoted their elution.

Indirect photometric detection provides another means for the determination of polyphosphates. They can be effectively displaced from the anion exchange resins by a powerful displacing ion such as the trimesate ion (a benzene tricarboxylate anion), and because they are relatively nonchromophoric they are detectable by the troughs they produce in the absorbance signal as they replace the UV-absorbing trimesate ions in the effluent (Figure 9.14).

With the advent of membrane suppressors that can cope with high concentrations of sodium hydroxide, suppressed conductometric detection has lately become a very practical means of detecting polyphosphate ions. Figure 9.15 shows a successful separation and detection of polyphosphates that employs a sodium hydroxide gradient in conjunction with a MicroMembrane suppressor.

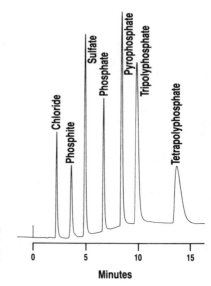

FIG. 9.15. Gradient elution of polyphosphate species. Eluent: 0.75 mM to 100 mM HCl. Detection: suppressed conductometric; Anion MicroMembrane suppressor. (From Ref. 7 with permission.)

9.4.2. Polyamines

Polyamines pose a problem for conductometric cation exchange systems much as polyphosphates do for anion exchange systems. The high charge of the polyprotonated species that form in the typical acidic cation eluents makes for difficult elution, which in turn complicates conductometric detection. Complexation of the amines, particularly with cupric ion, affords a simple means of facilitating both the separation and detection.

Figure 9.16 shows the separation of various mixtures of polyamines on a low-capacity cation exchanger. The mobile phase comprises two key ingredients, a small amount of cupric ion to complex the injected amines and a salt whose cations displace the cationic cupric–amine complexes from the resin. Copper–amine complexes as well as being intensely colored in the visible also absorb strongly in the UV, so a photometer monitoring at 254 nm is very effective in detecting the eluted complexes.

The copper complexes of the higher-molecular-weight amines, although they are probably only divalent, nevertheless have a high affinity for the resin phase probably because of the increasingly organophilic nature of the higher molecular weight complexes. In any event the higher affinity may be conveniently counteracted by increasing the displacing power of the eluent. This may be accomplished either by increasing its concentration or by substituting a polyvalent displacing ion such as calcium for a monovalent ion such as sodium. Since

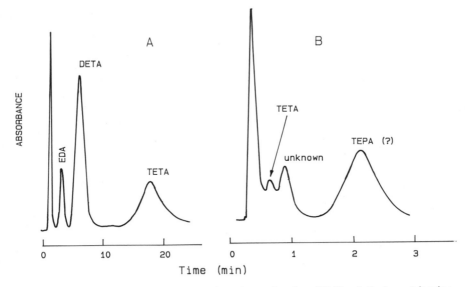

FIG. 9.16. Separation of polyamines [ethylene diamine (EDA), diethylene triamine (DETA), triethylene tetramine (TETA), and tetraethylene pentamine (TEPA)] as their copper complexes and detection by UV absorbance at 254 nm.[14] Stationary phase: surface sulfonated styrene DVB resin. Eluent for chromatogram A: 0.02 M CaCl$_2$, 0.01 M Na acetate, 10^{-4} M Cu(NO$_3$)$_2$. Eluent for chromatogram B: 0.2 M CaCl$_2$, 0.01 M Na acetate, 10^{-4} M Cu(NO$_3$)$_2$.

both sodium chloride and calcium chloride are non-UV absorbing at the monitoring wavelength there is ample latitude for making changes of this type.

9.4.3. Hydrophobic Anions

Many anions have very high affinities for anion exchange resins. Bulky organic substituents on the ions or low hydration energy are often cited as the reasons for their high affinity so they are often referred to as hydrophobic ions. Examples of hydrophobic anions are iodide, thiocyanate, the chlorinated acetic acids, surfactant anions in general, and many heavy metal cyanide complexes. Elution of such ions from the ion exchange resins commonly used for IC requires more concentrated eluents, which can complicate some conductometric methods of detection.

Ion exchange separation coupled with indirect photometric detection has been used effectively to analyze surfactant mixtures. Larsen and Pfeiffer[15,16] have used this approach to determine both anionic and cationic surfactants.

Mobile phase ion chromatography (MPIC), which uses ion interaction reagents in conjunction with suppressed conductometric detection, has proved to

TABLE 9.1. Reagents Suitable for Mobile Phase Ion Chromatography (MPIC) of Anions[a]

Analyte	Reagent	Acetonitrile concentration (%)
F^-, Cl^-, NO_2^-, etc.	$(C_4H_9)_4 N^+$	8
I^-, SCN^-	$(C_4H_9)_4 N^+$	15
Aromatic sulfonates	$(C_3H_7)_4 N^+$	20
Alkyl sulfonates, sulfates $<C_8$	$(C_4H_9)_4 N^+$	28
$Fe(CN)_6^{3-}$, $Au(CN)^{2-}$, etc.	$(C_4H_9)_4 N^+$	30
Alkyl sulfonates, sulfates $<C_8$	NH_4^+	30

[a]Reference 3.

be an extremely versatile solution to problems involving high-affinity ions. A typical system for anion analysis uses a surface active quaternary ammonium hydroxide as the IIR in conjunction with a macroporous polystyrene as stationary phase (Table 9.1). With very hydrophobic analytes such as surfactant anions, the ammonium ion serves adequately as the IIR. Elution of the analyte anions is promoted by using acetonitrile and sodium carbonate in the mobile phase; the acetonitrile modifies the adsorption of the IIR onto the hydrophobic substrate while the carbonate ion exchanges the analyte counterions from the IIR layer. A suppressor—packed bed or membrane based—removes the IIR and inorganic co-ions such as sodium and a conductivity cell completes the system.

Figures 9.17 and 9.18 show examples of MPIC applied to the determination

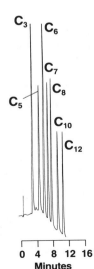

FIG. 9.17. Mobile phase ion chromatography (MPIC) of linear alkyl sulfonates. Stationary phase was a macroporous polystyrene-type polymer. Ion interaction reagent: 2.0 mM tetrabutylammonium hydroxide (TBAOH). Gradient of 18%–50% acetonitrile. Detection: suppressed conductometric. (Courtesy Dionex Corporation.)

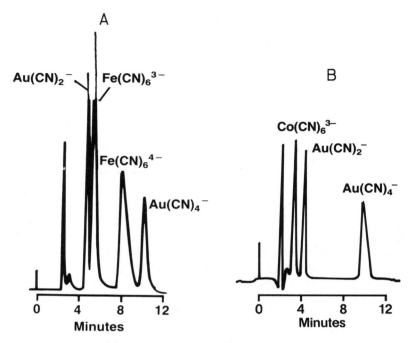

FIG. 9.18. Separation of metal cyanide complexes by MPIC. Eluent: 2 mM TBAOH, 0.02 M Na$_2$CO$_3$, 40% acetonitrile. Detection: suppressed conductometric. (From Ref. 3 with permission.)

of a variety of hydrophobic anions using tetra-butyl ammonium hydroxide as the IIR. MPIC has been extensively used for analyzing surfactants and for the determination of cyano-metal complexes in plating baths.

9.5. ANIONS OF VERY WEAK ACIDS

The anions of certain very weak acids can present problems for some of the common IC methods. Anions of very weak acids (pK greater than 7) cannot be determined by suppressed conductometric detection; anions such as carbonate, cyanide, sulfide, borate, and silicate, for example. The acids formed in the suppressor are only slightly dissociated and therefore too poorly conducting to be detectable. As an alternative to suppressed conductivity detection Pinschmidt[17] introduced a variation on conductivity detection that he called resistivity detection. The stationary phase is an anion exchanger. The eluent in this case is a salt of a strong acid, usually with some base added to make the species of interest sufficiently ionic as to be ion exchange active. The effluent from the separator passes to a bed of strong cation exchange resin in the hydrogen form just as it does in suppressed conductometric detection. But the effect of this bed is the opposite of what we normally observe. In the case of resistivity detection the conductance of the eluent is increased as the salt is converted to the more conducting acid and the conductance of the analyte is suppressed as it converts to a very weak acid in the second bed. Thus the weak acid anions appear as troughs in the highly conducting background of the strong acid. Figure 9.19 shows the detection of carbonate by resistivity detection.

Nonsuppressed conductometric methods using high pH eluents have been employed with success to measure weak acid anions. Silicate, for example, may be detected in a sodium hydroxide eluent as it displaces the more highly conducting hydroxide ion in the effluent (Figure 9.20). But for trace level determinations resistivity detection or nonsuppressed detection may lack the necessary sensitivity.

Electrochemical detection can be the best solution for some species.[19] Cyanide and sulfide are two anions that have been successfully determined using anion exchange separation followed by electrochemical detection (Figure 9.21). The relative insensitivity of the electrochemical detector to many ions in the sample can be a boon in solving severe matrix problems.

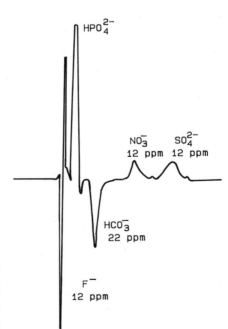

FIG. 9.19. Resistivity detection. Determination of carbonate and strong acid anions. Stationary phase: pellicular anion exchange resin. Eluent: 2.5 mN Na_3PO_4. Detection: nonsuppressed conductometric. (From Ref. 17 with permission.)

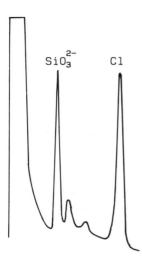

FIG. 9.20. Analysis of silicate. Stationary phase: low-capacity anion exchange resin. Eluent: 0.5 mM KOH. Detection: nonsuppressed conductometric. (From Ref. 18 with permission.)

Borate ion is a special case among weak acid anions in that we can take advantage of a unique feature of its chemistry to facilitate its detection. Boric acid is a very weak acid (pK 9.2), but in the presence of certain polyhydric alcohols and sugars it forms complexes that are stronger acids than the boric acid itself. The acid strengths of some of these complexes are listed in Table 9.2.

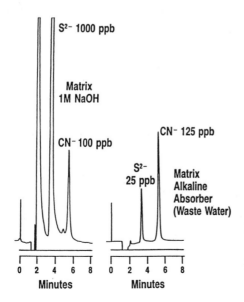

FIG. 9.21. Determination of cyanide and sulfide in strongly alkaline solutions. Stationary phase: pellicular anion exchange resin. Eluent: 0.1 M NaOH, 0.5 M Na acetate, 0.5% ethylene diamine. Detection: electrochemical. (Courtesy Dionex Corporation.)

TABLE 9.2. Acid Dissociation
Constants for Various Complexes
of Boric Acid with Polyhydric
Alcohols[a]

Complex	pK_a
Boric acid	9.2
Boric acid + mannitol (1 molar)	4
Boric acid + fructose (1 molar)	4
Boric acid + glycerol (1 molar)	6.5
Boric acid + glycerol (3.5 molar)	5.7

[a]Reference 20.

Since the addition of enough mannitol, for example, can increase the apparent acidity of boric acid by a factor of roughly 10^5, the addition of mannitol to the eluent in an ion exclusion type separation was found to be a very effective means of enhancing its conductometric detectability.

One practical application of this chromatographic scheme was the determination of occluded borate impurities in magnesium hydroxide slurries. Since chloride and fluoride were also present in trace amounts, a chromatographic separation was necessary. A feature of the total procedure was the manner in which the occluded impurities were released into solution. Magnesium hydroxide was dissolved simply by stirring the slurry with a small quantity of strong cation exchanger in the hydrogen form:

$$2R-SO_3^- \; H^+ \; + \; Mg(OH)_2 \rightarrow (R-SO_3^-)_2Mg^{2+} \; + \; H_2O$$

"Solid–solid" reactions of this type were first proposed many years ago by Osborn,[21] who made an extensive study of the dissolution of precipitates by ion exchange resins. In this instance it is an extremely effective way of dissolving the solid and releasing occluded impurities into solution without adding a large amount of another anion as would happen if a simple acid was used. The separation–detection system comprised a strong cation exchanger in the hydrogen form as stationary phase, a solution of mannitol as mobile phase, and a conductivity detector.

Figure 9.22 shows the ion exclusion separation of three anion impurities as their acids. Chloride of course elutes first since it is excluded completely from the resin. Hyrofluoric acid is somewhat retarded and separated from the HCl band. The boric acid–mannitol complex is completely separated from these interfering species and easily detected by its conductance. In the absence of mannitol no peak for boric acid is observed.

Carbonate has been successfully determined by ion exclusion. Using hydrogen form cation exchanger as stationary phase and pure water as the mobile

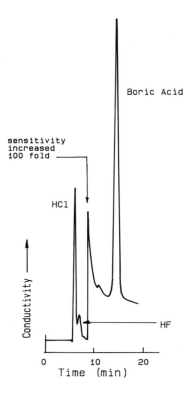

Boric Acid

sensitivity
increased
100 fold

HCl

Conductivity ⟶

HF

Time (min)
0   10   20

FIG. 9.22. Ion exclusion separation of boric acid mannitol complex from HF and HCl.[22] Stationary phase: fully sulfonated polystyrene DVB(4%) resin. Eluent: 0.2 M mannitol in water. Detection: nonsuppressed conductometric.

phase, strong acids elute at the column void volume while the weak carbonic acid is retarded. Carbonic acid is sufficiently ionized in the deionized water environment to be measurable by a conductance detector.[23]

9.6. ION EXCLUSION

The analytical potential of ion exclusion has been recognized for some time,[24] but early workers were hampered by the primitive state of detection. Now ion exclusion has become a widely practiced IC technique largely because of recent developments in automated detection.

Richards[25] was the first to adapt ion exclusion to modern LC practice. Building on earlier work, Richards coupled the separation to a UV monitor and demonstrated rapid, efficient separation and detection of a number of unsaturated organic acids. Turkelson and Richards[26] later extended this approach to the determination of acids of the Krebs cycle (see also Chapter 5, Section 5.2.4).

Not all weak acids are accessible to UV monitoring. In the late 1970s workers at Dionex Corporation[27–29] pioneered the application of conductometric methods to ion exclusion, thus opening it up to a wider range of applications. The major problem they confronted was the background of mineral acid in the mobile phase. This was not a problem for UV detection since it was essentially invisible, but it tended to mask the weak acids in a conductance mode. The solution devised by Rich and co-workers employed dilute hydrochloric acid as the mobile phase followed by stripping of the mineral acid from the effluent in a suppressor bed. The suppressor was essentially the same resin as the separator, a strong cation exchange resin, but in the silver ion form. The hydrochloric acid was simply removed by ion exchange and precipitation in this second bed:

$$R-SO_3^- \ Ag^+ + HCl \rightarrow R-SO_3^- \ H^+ + AgCl \ \text{(precipitated on resin)}$$

Many successful schemes were devised around this system.

Although the silver suppressor worked well it was not without problems. Mainly it was difficult and somewhat inconvenient to regenerate since the accumulated precipitate of silver chloride had to be removed. More recent ion exclusion systems use a quite different approach to suppression. A strong organic acid such as perfluoro butyric acid or octane sulfonic acid is added to the mobile phase to promote association of the weak acids. Following the separation this acid is not completely removed but instead the hydronium ion is exchanged in a membrane reactor for a low-mobility cation such as the tetrabutylammonium ion.

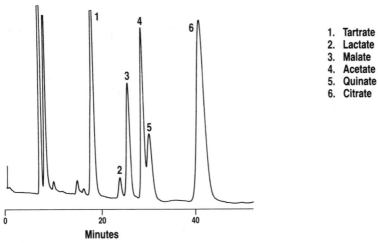

1. Tartrate
2. Lactate
3. Malate
4. Acetate
5. Quinate
6. Citrate

FIG. 9.23. Determination of organic acids in coffee. Stationary phase: fully sulfonated polystyrene DVB resin. Eluent: perfluorobutyric acid pH 2.8. Detection: Suppressed conductometric, Anion MicroMembrane suppressor using tetrabutylammonium hydroxide as regenerant. (Courtesy Dionex Corporation.)

This salt of a large cation and large anion is relatively poorly conducting and permits easy conductometric detection of the organic acids against this low background.[29] Figure 9.23 shows a typical separation of organic acids using this type of system.

Combinations of ion exchange and ion exclusion separation are effective when it is necessary to determine inorganic as well as organic ions in the same sample.

9.7. AMINO ACIDS

Since the early 1950s ion exchange chromatography has had an important place in the analytical chemistry of amino acids. In light of their importance and their unique acid–base chemistry it is instructive to examine their ion exchange chromatographic behavior. A brief introduction to their acid–base chemistry is helpful.

9.7.1. Acid–Base Chemistry of Alpha-Amino Acids (See Also Appendixes E and F)

Although there are a great number of amino acids we will be concerned here with the twenty alpha-amino acids commonly found in proteins (Appendix E). With the exception of proline all have in common a free carboxyl group and a free unsubstituted amino group attached in the alpha position to a central carbon atom (Figure 9.24). The feature that distinguishes one member from another within this group is the side chain R. Depending on the nature of R, the group can be subdivided into neutral, acidic, and basic acids. Neutral acids are those where R is neutral—glycine, tyrosine, and threonine, for example. This subgroup has been further classified into those acids that are nonpolar and those that are polar.[30]

The acidic amino acids, aspartic and glutamic, have a carboxylic acid group in their side chains. The basic amino acids, arginine, lysine, and histidine, have an amino group in the side chain.

A   B

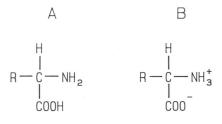

FIG. 9.24. The general structural formula for the common amino acids. (A) The undissociated form. (B) The zwitterionic form.

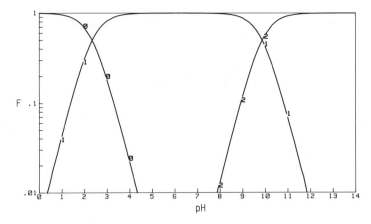

FIG. 9.25. Logarithmic diagram for alanine. (See Appendix A for explanation of logarithmic diagrams.)

The bipolar form or zwitterion of Figure 9.24B is the accepted structure of the amino acids either in the solid or when they are dissolved in water. This structure is consistent with the relatively high melting points of the solid amino acids and their high dipole moments in solution. The position of the proton in the structure is also consistent with the relative basicities of the amino and carboxyl groups.

Knowing the disposition of charge of the amino acids as a function of pH is important to an understanding of their ion exchange behavior, so in this connection we will examine alanine, aspartic acid, and lysine as exemplifying the three types of acid.

At a low pH (less than 1) alanine exists almost entirely as a dibasic acid, $H_3N^+CH(CH_3)COOH$, with a single positive charge. As the pH is increased, the carboxylic acid, the stronger acid of the two, dissociates first and the developing negative charge of the freed carboxyl group gradually neutralizes the positive charge of the protonated amino group. The growth and disappearance of various forms is conveniently presented in logarithmic diagrams* such as Figures 9.25–9.27. Curve 0 of Figure 9.25 illustrates the decline in the concentration of the undissociated carboxylic acid group while curve 1 represents the build-up of free carboxylate. Though it does not show in the figure because of the truncation of the data, it turns out that the carboxylic acid is almost completely converted to carboxylate at a pH of about 6, at which point the charge of the carboxylate exactly balances the positive charge of the ammonium group and the molecule is neutral. This unique pH is the isoelectric pH or isoelectric point of the acid. At

*See Appendix A.

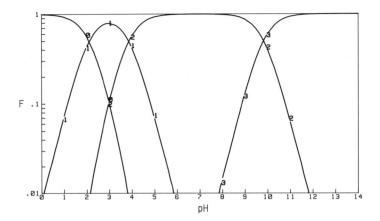

FIG. 9.26. Logarithmic diagram for aspartic acid.

this pH the overall net charge of the species is zero, which is supported by the fact that it will not migrate in an electric field.

Aspartic acid at low pH is a tribasic acid with a single positive charge, $NH_3^+ CH(CH_2COOH)COOH$. Counterbalancing of this charge starts as the pH is raised and the first carboxylic group is neutralized (Figure 9.26). However, before this group is completely converted to carboxylate, neutralization of the second carboxylic group begins. At a pH of roughly 3 the fractions of mono-carboxylated and di-carboxylated forms are 0.8 and 0.1, respectively, enough total equivalents of carboxylate to just balance the one equivalent of protonated

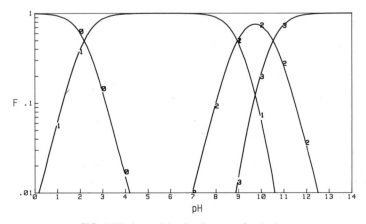

FIG. 9.27. Logarithmic diagram for lysine.

amino group. Thus, the isoelectric point of aspartic is around pH 3—likewise for glutamic acid.

A basic acid such as lysine on the other hand, which contains two amino groups and one carboxylic group, is doubly positive at low pH. Total neutralization of the carboxylic acid group produces a mono-positive species and it is not until some neutralization of the protonated amino groups has taken place that the zwitterionic form is produced. The isoelectric point is reached when the total equivalents of residual protonated amino forms together balance the one equivalent of carboxylate group. From Figure 9.27 it is evident that this occurs between pH 9 and 10—9.74 to be precise.

Since the isoelectric point of an amino acid represents a point of neutrality, it is inviting to conclude that it also represents a point of minimum ion exchange activity, a point that is where the amino acid should be disinclined to attach to either a cation or anion exchanger. This is not necessarily the case, as the following argument will illustrate.

We saw in Section 4.2.6 of Chapter 4 that the pH of the external phase can be considerably higher or lower than the pH of the resin phase. It is quite conceivable, therefore, that the pH of the external phase could be on one side of the isoelectric point and the pH of the resin phase be on the other. So while the conditions in the external phase might suggest rejection (or retention) of the amino acid, the pH of the resin phase would dictate quite the opposite. It is important to bear this in mind in attempting to understand the ion exchange behavior of amino acids.

9.7.2. Ion Exchange Separation of Amino Acids

Amino acids can be separated on either cation or anion exchangers. In a typical cation separation, the mixture of amino acids is loaded onto the acid form of a cation exchanger. The acids are then selectively displaced from the column by essentially two influences:

1. Ion exchange displacement of the cationic forms of the acids by cations in the eluent; sodium or lithium are commonly used.
2. Step or gradient increases in the pH of the eluent to render all the acids less positive, and therefore more elutable.

Just the opposite conditions apply to anion exchange separation—loading at a relatively high pH and decreasing the pH either progressively or by steps.

Figures 9.28 and 9.29 provide a comparison of a cation and anion separation of amino acids. It is apparent that acid–base properties play some role in determining the order of elution but it is not the only factor.

For the cation separation, the basic amino acids arginine, histidine, and lysine are di-positive in charge at the pH of the eluents used, and since the resin

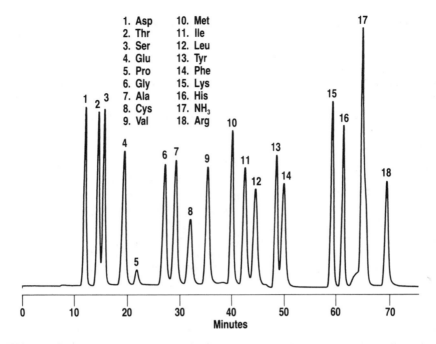

FIG. 9.28. Separation of an amino acid hydrolysate standard on a low-capacity cation exchange resin. Eluent: buffers of pH 3.24, 7.11, and 11.0 introduced in that order. Detection: photometric after postcolumn reaction with ninhydrin reagent. (From Ref. 31 with permission.)

is considerably more acidic than the aqueous phase these acids are strongly retained and the order of their elution is probably due to polarity differences. The term polarity in this context implies the collective hydrophobic and hydration effects discussed in Section 4.2.8.

The order of elution of neutral acids cannot be explained entirely on the basis of charge phenomena. For example, the acid–base properties of glycine, alanine, valine, and leucine are too alike to be solely responsible for the separation or order of elution of these acids. However, a factor that does correlate with their ion exchange behavior is their relative polarity. Certainly the elution order of glycine, alanine, valine, and iso-leucine is reflected in their hydrophobicities if one accepts that bulkiness of the R group is a measure of hydrophobicity.

Similar arguments can be applied to the retention behavior of the other neutral acids as well. The relatively high polarity of threonine and serine explains their early elution while the aromatic R groups of tryptophan and phenylalanine are the probable cause of their relatively long retentions.

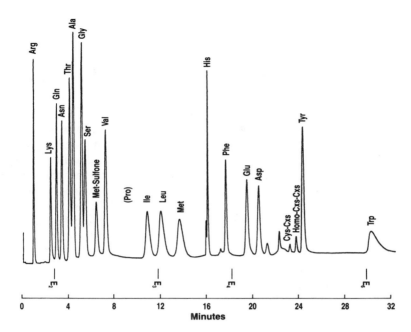

FIG. 9.29. Anion exchange separation of amino acids. Stationary phase: pellicular anion exchange resin. Eluents: E_1, 28 mM NaOH, 8 mM Na$_2$B$_4$O$_7$; E_2, 70 mM NaOH, 20 mM Na$_2$B$_4$O$_7$; E_3, 160 mM NaOAc; E_4, 320 mM NaOAc. Detection: photometric after postcolumn reaction with o-phthalaldehyde reagent. (From Ref. 31 with permission.)

The early elution of aspartic and glutamic acids can be attributed to their relatively low isoelectric points especially since the pH of the most acidic eluent is close to the isoelectric pH. The acidity of the resin phase, however, is likely to be much greater than the external phase, thus favoring the cationic form of these two basic acids. The fact that they are retained at all supports an argument that the resin phase pH is probably below their isoelectric points. Notable too is the elution of threonine and serine before glutamic acid even though they have considerably higher isoelectric points than glutamic acid. The order of elution of aspartic and glutamic acids, so alike in acid–base character, is best explained by their polarity differences.

Similar arguments can be proposed to explain the order of retention in an anion exchange separation of amino acids (Figure 9.29). In this case, glutamic and aspartic acids are the most retarded since they exist as di-negative anions at the prevailing pHs of the eluents used. Polarity differences again seem to account best for the order of elution of the neutral acids, although there are exceptions to this—the order of glycine and alanine for example. The exceptionally long retention of tyrosine with its phenolic R group correlates with the well-known high affinity of such groups for styrene-based anion exchange resins.

With the exception of histidine, the basic acids elute earliest, indicating a contribution from acid–base character. Histidine is exceptional, probably because its isoelectric point is much lower than those of the other basic acids and it retains its anionic character to much lower pH levels.

The measurement of the amino acids is a good example of postcolumn reaction followed by photometric detection. In the cation exchange separation, ninhydrin was the reagent added, followed by colorimetric measurement of the intensely blue complex. In the case of the anion exchange separation a postseparation addition of o-phthalaldehyde, which forms a fluorescing complex with the amino acids, is the basis of this very sensitive fluorimetric method.

In summary, the analysis of amino acids exemplifies a particularly challenging problem for IC: it is a difficult separation problem; the order of elution presents a challenge to our understanding of resin selectivity and the part played by pH effects in the ion exchange process, with much still to be learned in this regard; it is a difficult detection problem that has yielded to few but nonetheless very effective solutions.

9.8. CARBOHYDRATES

Carbohydrates are another class of compounds whose ion exchange behavior is profoundly affected by the pH of the mobile phase. Not usually considered to be ionic, carbohydrates and other polyhydroxy compounds are nevertheless acidic enough (Table 9.3) to become appreciably ionized in strongly basic solutions. They can then participate in ion exchange reactions with strongly basic resins. Rocklin and Pohl[33] described the separation of several sugar alcohols, saccharides, and oligosaccharides by anion exchange IC and their sensitive detection by the pulsed amperometric detector. They examined the effects on k-prime and resolution of eluent strength, eluting ion, and the temperature of the separation. Detection limits as low as 30 ppb were attainable for some compounds. Figure 9.30 and 9.31 represent some of the separations that are possible with this remarkable advance in ion chromatography.

Note. For the reader who wishes to keep up to date on the manifold applications of ion chromatography, the publication *Ion Chromatography* in the Chemical Abstracts Selects Series published by the Chemical Abstracting Service of the American Chemical Society is highly recommended.

TABLE 9.3. Ionization
Constants (Hydroxyl Group)
of Carbohydrates in Water[a]

Compound	pK_a
D-Glucose	12.35
D-Galactose	12.35
D-Mannose	12.08
D-Arabinose	12.43
D-Ribose	12.21
D-Xylose	12.29
2-Deoxy-D-glucose	12.52
2-Deoxy-D-ribose	12.65
Lactose	11.98
Maltose	11.94
Raffinose	12.74
Sucrose	12.51
D-Fructose	12.03
D-Glucitol	13.57
D-Mannitol	13.50
Glycerol	14.4

[a]Reference 32.

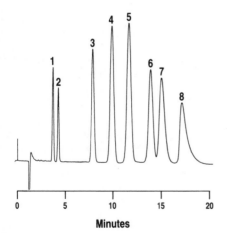

Sugar	μg/mL
1. Fucose	25
2. Deoxyribose	25
3. Arabinose	25
4. Galactose	25
5. Glucose	25
6. Xylose	25
7. Mannose	50
8. Fructose	50

FIG. 9.30. Separation of common monosaccharides by anion exchange and pulsed amperometric detection. (Courtesy Dionex Corporation.)

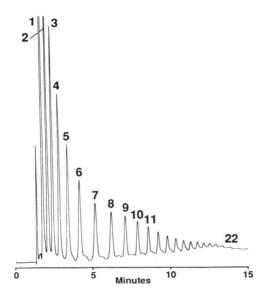

FIG. 9.31. Separation of oligo- and polysaccharides on an anion exchange resin using 0.15 *M* NaOH as eluent and pulsed amperometric detection. (Courtesy Dionex Corporation.)

REFERENCES

1. H. Small, Ion Chromatography in Trace Analysis, in *Trace Analysis*, Vol. 1 (J. F. Lawrence, ed.), Academic Press, New York (1981).
2. E. Sawicki, J. D. Mulik, and E. Wittgenstein, *Ion Chromatographic Analysis of Environmental Pollutants*, Vols. 1 and 2, Ann Arbor Science Publishers, Ann Arbor, Michigan (1978 and 1979).
3. E. L. Johnson and K. K. Haak, Anion Analysis by Ion Chromatography, in *Liquid Chromatography in Environmental Analysis* (J. F. Lawrence, ed.). The Humana Press, Clifton, New Jersey (1983).
4. D. T. Gjerde and J. S. Fritz, *Ion Chromatography*, 2nd edition, Huethig Publishing, Mamaroneck, New York (1987).
5. H. Small and J. Solc, Ion Chromatography—Principles and Applications, in *The Theory and Practice of Ion Exchange* (M. Streat, ed.), The Society of Chemical Industry, London (1976).
6. Dionex Corporation, *Ion Chromatography Cookbook*. Dionex Corporation, Sunnyvale, California (1987).
7. R. D. Rocklin, C. A. Pohl, and J. A. Schibler, Gradient Elution in Ion Chromatography, *J. Chromatogr.* **411**, 107–119 (1987).
8. R. E. Barron and J. S. Fritz, Reproducible Preparation of Low-Capacity Anion-Exchange Resins, *React. Polym.* **1**, 215–226 (1983).
9. D. T. Gjerde, J. S. Fritz, and G. Schmuckler, Anion Chromatography with Low-Conductivity Eluents, *J. Chromatogr.* **186**, 509–519 (1979).
10. Dionex Corporation, *Ion Chromatography Cookbook*, Dionex Corporation, Sunnyvale, California (1987).
11. H. Small, unpublished results.

12. Dionex Corporation, Determination of Sequestering Agents, Application Note No. 44, Dionex Corporation, Sunnyvale, California (1982).

13. H. Small and T. E. Miller, Jr., Indirect Photometric Chromatography, *Anal Chem.* **54**, 462–469 (1982).

14. H. Small, unpublished results.

15. J. R. Larson and C. D. Pfeiffer, Determination of Organic Ionic Compounds by Liquid Chromatography with Indirect Photometric Detection, *J. Chromatogr.* **259**, 519–521 (1983).

16. J. R. Larson and C. D. Pfeiffer, Determination of Alkyl Quaternary Ammonium Compounds by Liquid Chromatography with Indirect Photometric Detection, *Anal. Chem.* **55**, 393–396 (1983).

17. R. K. Pinschmidt, Ion Chromatographic Analysis of Weak Acid Ions, in *Ion Chromatographic Analysis of Environmental Pollutants,* Vol. 2 (J. D. Mulik and E. Sawicki, eds.), Ann Arbor Science Publishers, Ann Arbor, Michigan (1979).

18. Waters Chromatography Division of Millipore Corporation, Analysis of Silicate at Trace Levels, Application Brief No. 2004, Waters Division of Millipore Corporation, Milford, Massachusetts.

19. R. D. Rocklin and E. L. Johnson, Determination of Cyanide, Sulfide, Iodide, and Bromide by Ion Chromatography with Electrochemical Detection, *Anal. Chem.* **55**, 4–7 (1983).

20. A. A. Nemodruk and Z. K. Karalova, *Analytical Chemistry of Boron,* p. 39, Humphrey Science Publishers, Ann Arbor, Michigan (1969).

21. G. H. Osborn, Ion-Exchange Resins in Analytical Chemistry. Application of Ion-Exchange Resins to the Analysis of Insoluble Substances, *Analyst* **78**, 220–221 (1953).

22. H. Small, unpublished results.

23. K. Tanaka and J. S. Fritz, Determination of Bicarbonate by Ion-Exclusion Chromatography with Ion-Exchange Enhancement of Conductivity Detection, *Anal. Chem.* **59**, 708–712 (1987).

24. P. Jandera and J. Churacek, Ion-Exchange Chromatography of Carboxylic Acids, *J. Chromatogr.* **86**, 351–421 (1973).

25. M. Richards, Separation of Mono- and Dicarboxylic Acids by Liquid Chromatography, *J. Chromatogr.* **115**, 259–261 (1975).

26. V. T. Turkelson and M. Richards, Separation of the Citric Acid Cycle Acids by Liquid Chromatography, *Anal. Chem.* **50**, 1420–1423 (1978).

27. W. Rich, F. Smith Jr., L. McNeil, and T. Sidebottom, Ion Exclusion Coupled to Ion Chromatography, in *Ion Chromatographic Analysis of Environmental Pollutants,* Vol. 2 (J. D. Mulik and E. Sawicki, eds.), pp. 17–30, Ann Arbor Science Publishers, Ann Arbor, Michigan (1979).

28. W. Rich, E. L. Johnson, L. Lois, P. Kabra, B. Stafford, and L. Marton, Determination of Organic Acids in Biological Fluids by Ion Chromatography: Plasma Lactate and Pyruvate and Urinary Vannllylmandelic Acid, *Clin. Chem.* **26**(10), 1492–1498 (1980).

29. C. A. Pohl, R. W. Slingsby, E. L. Johnson, and L. Angers, Method and Apparatus for Ion Analysis and Detection Using Reverse Mode Suppression, U.S. Patent No. 4,455,233 (1984).

30. A. L. Lehninger, *Biochemistry*, 2nd edition, Chapter 4, Worth Publishers, New York (1975).

31. Dionex Corporation, *BioLC Series Amino Acid Analysis System,* Dionex Corporation, Sunnyvale, California (1988).

32. J. A. Rendleman, Ionization of Carbohydrates in the Presence of Metal Hydroxides and Oxides, *Carbohydrates in Solution,* Advances in Chemistry Series No. 117, p. 51, American Chemical Society, Washington, D.C. (1973).

33. R. D. Rocklin and C. A. Pohl, Determination of Carbohydrates by Anion Exchange Chromatography with Pulsed Amperometric Detection, *J. Liq. Chromatogr.* **6**(9), 1577–1590 (1983).

Appendixes

Appendix A

Logarithmic Diagrams

Logarithmic diagrams are a convenient way of presenting information on complex ionic equilibria.[1,2] For example, dissociation of a simple monoprotic acid HA in aqueous solution,

$$HA \rightleftharpoons H^+ + A^-$$ (A.1)

yields species HA and A^-. A straightforward calculation gives the fractions of acid present as each species as a function of $[H^+]$:

$$F_{HA} = 1/\{(K/[H^+]) + 1\}$$ (A.2)

$$F_{A^-} = 1/\{1 + ([H^+]/K)\}$$ (A.3)

where K is the acid dissociation constant, defined as

$$K = [H^+][A^+]/[HA]$$ (A.4)

Equations (A.2) and (A.3) can be displayed in a number of ways. A plot of F versus pH is called a distribution diagram; a plot of log F versus pH is called a logarithmic diagram. The latter method has been used in Figures A.1–A.21 to display information on a number of species of importance to ion chromatography. Logarithmic diagrams for amino acids that are representative of their various classes are presented in Chapter 9, Section 9.7.1. Histidine has also been included (Figure A.9) as it is distinctly different from the other basic amino acids in its dissociation. Acid–base dissociation constants used to construct these diagrams were obtained from Ref. 3.

The curves are identified in the following way. The curve labeled 0 (zero) gives the abundance of the most protonated species; for example, for phosphoric acid it would refer to the H_3PO_4 species, for alanine to the $NH_3^+ CH(CH_3)COOH$ species. Curve 1 applies to the next most protonated species ($H_2PO_4^-$, $NH_3^+ CH(CH_3)COO^-$), curve 2 to the next, and so on.

For the reader who might wish to construct such diagrams for other equilibria, the relationships for dibasic and tribasic acids are as follows. For a dibasic acid,

$$F_{H_2A} = 1/\{1 + (K_1/[H^+]) + (K_1K_2/[H^+]^2)\}$$

$$F_{HA^-} = 1/\{([H^+]/K_1) + 1 + (K_2/[H^+])\}$$

$$F_{A^{2-}} = 1/\{([H^+]^2/K_1K_2) + ([H^+]/K_2) + 1\}$$

where K_1 and K_2 are the acid dissociation constants for the equilibria

$$H_2A \rightleftharpoons HA^- + H^+$$

and

$$HA^- \rightleftharpoons A^{2-} + H^+$$

respectively. For a tribasic acid,

$$F_{H_3A} = 1/\{1 + (K_1/[H^+]) + (K_1K_2/[H^+]^2) + (K_1K_2K_3/[H^+]^3)\}$$

$$F_{H_2A^-} = 1/\{([H^+]/K_1) + 1 + (K_2/[H^+]) + (K_2K_3[H^+]^2)\}$$

$$F_{HA^{2-}} = 1/\{([H^+]^2/K_1K_2) + ([H^+]/K_2) + 1 + (K_3[H^+])\}$$

$$F_{A^{3-}} = 1/\{([H^+]^3/K_1K_2K_3) + ([H^+]^2/K_2K_3) + ([H^+]/K_3) + 1\}$$

where again K_1, K_2, and K_3 are the acid dissociation constants.

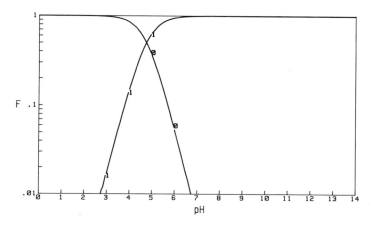

FIG. A.1. Logarithmic diagram for acetic acid.

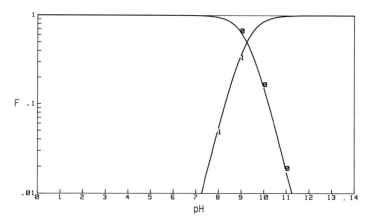

FIG. A.2. Logarithmic diagram for ammonium ion.

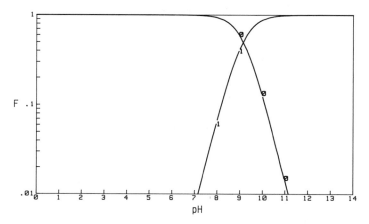

FIG. A.3. Logarithmic diagram for o-boric acid.

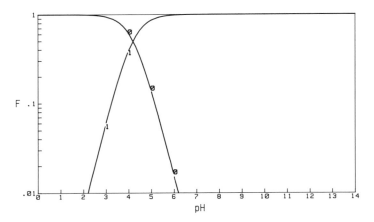

FIG. A.4. Logarithmic diagram for benzoic acid.

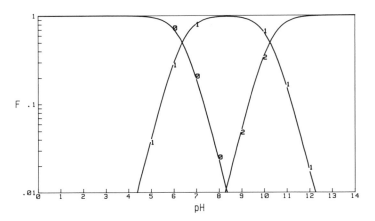

FIG. A.5. Logarithmic diagram for carbonic acid.

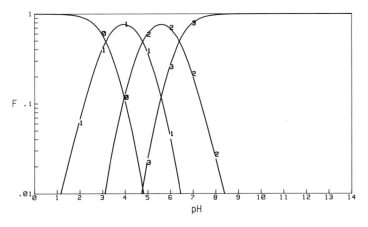

FIG. A.6. Logarithmic diagram for citric acid.

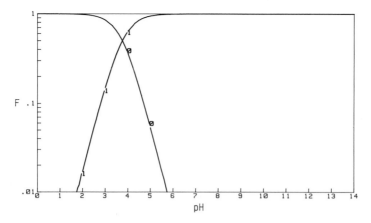

FIG. A.7. Logarithmic diagram for formic acid.

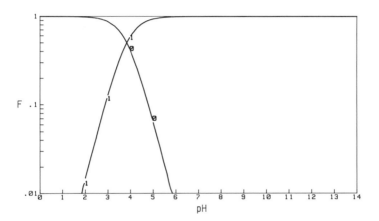

FIG. A.8. Logarithmic diagram for glycolic acid.

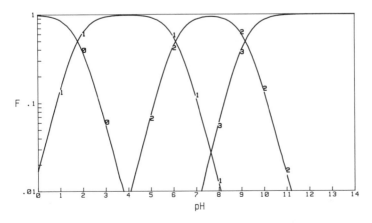

FIG. A.9. Logarithmic diagram for histidine.

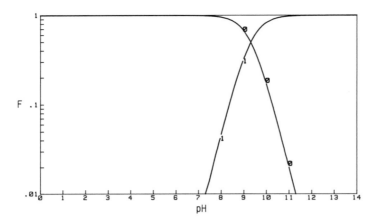

FIG. A.10. Logarithmic diagram for hydrocyanic acid.

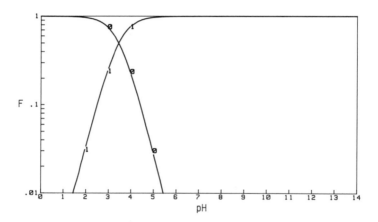

FIG. A.11. Logarithmic diagram for hydrofluoric acid.

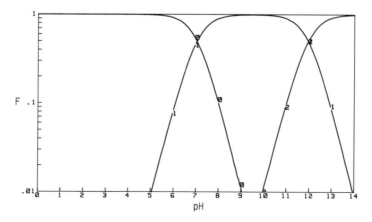

FIG. A.12. Logarithmic diagram for hydrogen sulfide.

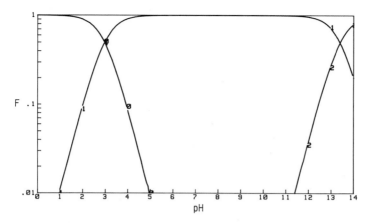

FIG. A.13. Logarithmic diagram for o-hydroxybenzoic acid.

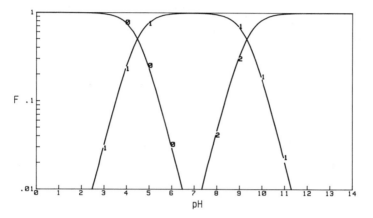

FIG. A.14. Logarithmic diagram for p-hydroxybenzoic acid.

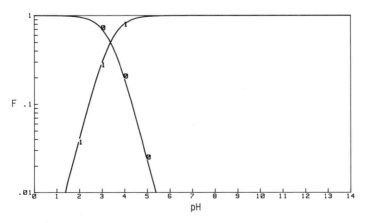

FIG. A.15. Logarithmic diagram for nitrous acid.

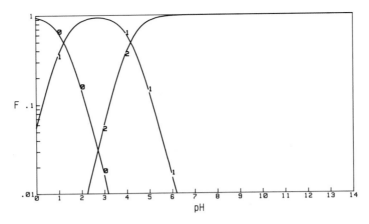

FIG. A.16. Logarithmic diagram for oxalic acid.

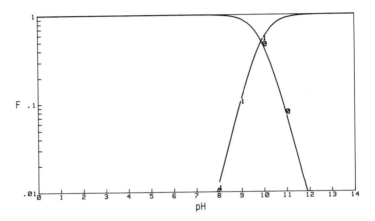

FIG. A.17. Logarithmic diagram for phenol.

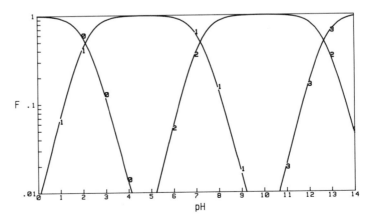

FIG. A.18. Logarithmic diagram for o-phosphoric acid.

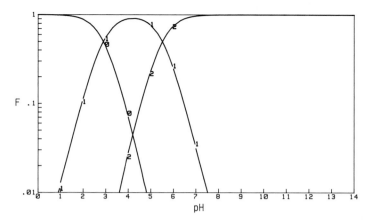

FIG. A.19. Logarithmic diagram for o-phthalic acid.

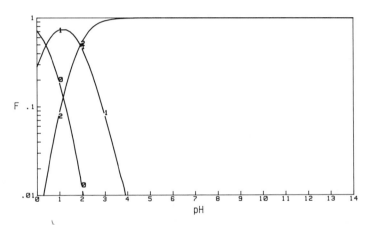

FIG. A.20. Logarithmic diagram for sulfuric acid.

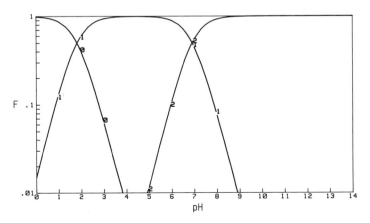

FIG. A.21. Logarithmic diagram for sulfurous acid.

Appendix B

Limiting Equivalent Conductivities of Ions at 25°C[4]

Ion	λ_o	Ion	λ_o
H^+	349.8	OH^-	198.6
Li^+	38.6	F^-	55.4
Na^+	50.1	Cl^-	76.35
K^+	73.5	Br^-	78.14
Rb^+	77.8	I^-	76.8
Cs^+	77.2	N_3^-	69
Ag^+	61.9	NO_3^-	71.46
Tl^+	74.7	ClO_3^-	64.6
NH_4^+	73.5	BrO_3^-	55.7
NMe_4^+	44.9	ClO_4^-	67.3
NEt_4^+	32.6	IO_4^-	54.5
NPr_4^+	23.4	HCO_3^-	44.5
NBu_4^+	19.4	Formate	54.5
NAm_4^+	17.4	Acetate	40.9
Mg^{2+}	53.0	Propionate	35.8
Ca^{2+}	59.5	Benzoate	32.3
Sr^{2+}	59.4	SO_4^{2-}	80.0
Ba^{2+}	63.6	Oxalate	74.1
Cu^{2+}	56.6	CO_3^{2-}	69.3
Zn^{2+}	52.8	$Fe(CN)_6^{3-}$	100.9
La^{3+}	69.7	$Fe(CN)_6^{4-}$	110

Appendix C

Conductance of Carbonic Acid as It Is Neutralized by Sodium Hydroxide

Certain behavior in ion chromatography (Chapter 5, Section 5.3 and Chapter 7, Section 7.3.4d) can be explained by considering the conductance changes that occur in the very early stages of neutralization of carbonic acid by a base. The following derivation yields an expression for the specific conductance, κ, of carbonic acid partially neutralized by sodium hydroxide.

When carbonic acid is partially neutralized by sodium hydroxide the system contains the species Na^+, H^+, H_2CO_3, HCO_3^-, and CO_3^{2-}. Since it is only the very early stages of neutralization that are being considered, it can be assumed that the concentration of the carbonate ion is negligible, and it follows then that

$$[Na^+] + [H^+] = [HCO_3^-] \tag{C.1}$$

The specific conductance of the partially neutralized carbonic acid is then given by

$$\kappa = \lambda_{H^+} [H^+] + \lambda_{Na^+} [Na^+] + \lambda_{HCO_3^-} [HCO_3^-] \tag{C.2}$$

or substituting from (C.1)

$$\kappa = \lambda_{H^+} [H^+] + \lambda_{Na^+} [Na^+] + \lambda_{HCO_3^-} ([Na^+] + [H^+]) \tag{C.3}$$

Since limiting conductances are known (Appendix B) and $[Na^+]$ is obtained from the amount of sodium hydroxide added, in order to calculate κ it only remains to develop an expression for $[H^+]$ in terms of known quantities. The acid–base equilibrium for the first dissociation of carbonic acid is defined by

$$K_1 = [H^+][HCO_3^-]/[H_2CO_3] = 4.3 \times 10^{-7} \tag{C.4}$$

Substituting from (C.1) it follows that

$$[H^+]([Na^+] + [H^+]) = K_1 [H_2CO_3] \tag{C.5}$$

Again, since the degree of neutralization is small, the concentration of carbonic acid throughout this region will remain essentially constant and close to the initial concentration of carbonic acid denoted C. Therefore,

$$[H^+]([Na^+] + [H^+]) = K_1C \qquad (C.6)$$

or

$$[H^+]^2 + [Na^+][H^+] - K_1C = 0 \qquad (C.7)$$

Substituting various values for $[Na^+]$, the quadratic (C.7) can then be easily solved for a positive root of $[H^+]$ thus providing a value for $[H^+]$ in equation (C.3). Figure 7.9 expresses the results obtained from this calculation assuming an initial concentration of 0.005 M carbonic acid.

Appendix D

Conductance of a Strong Acid in a Background of Weak Acid

This calculation relates to the conductance of the acid from a sample anion S^- in a weak acid effluent from a suppressor (Chapter 7, Section 7.3.8).

Assume that the weak acid, HA, has a background concentration C_A. The concentrations of hydronium ions and analyte anions from the sample species, HS, are respectively $[H^+]$ and $[S^-]$. (It is assumed in this calculation that HS is a strong acid.)

The background acid contributes hydronium and anions at concentrations $[H^+]_A$ and $[A^-]$, respectively. So the total hydronium ion concentration, denoted $[H^+]_{tot}$, is given by

$$[H^+]_{tot} = [H^+]_A + [H^+]_S \qquad (D.1)$$

From the acid–base equilibrium expression

$$[H^+][A^-]/[HA] = K \qquad (D.2)$$

where K is the acid dissociation constant, it follows that

$$([H^+]_A + [H^+]_S) [A^-]/[HA] = K \qquad (D.3)$$

Since the concentration of cations must equal that of the anions, then

$$[H^+]_{tot} = [A^-] + [S^-]$$

i.e.,

$$[A^-] = [H^+]_{tot} - [S^-]$$

also

$$[HA] = C_A - [A^-]$$

Substituting in (D.3) above gives

$$[H^+]_{tot} ([H^+]_{tot} - [S^-]) = K(C_A - [A^-]) = K(C_A - [H^+]_{tot} + [S^-])$$

$$[H^+]_{tot}^2 - [S^-][H^+]_{tot} = KC_A - K[H^+]_{tot} + K[S^-]$$

or

$$[H^+]_{tot}^2 + [H^+]_{tot}(K - [S^-]) - K(C_A + [S^-]) = 0 \qquad (D.4)$$

Solving the above quadratic for a positive root of $[H^+]_{tot}$ enables the calculation of the conductance of the system from the following expressions:

$$\kappa = 1000\,(\lambda_{H^+}[H^+]_{tot} + \lambda_{A^-}[A^-] + \lambda_{S^-}[S^-])$$

$$= 1000\,\{\lambda_{H^+}[H^+]_{tot} + \lambda_{A^-}([H^+]_{tot} - [S^-]) + \lambda_{S^-}[S^-]\}$$

The results of this calculation for low concentrations of HCl in 0.005 M carbonic acid are expressed in Figures 7.14 and 7.15.

Appendix E

Ionization Constants and pH Values at the Isoelectric Points (pI) of the Common Amino Acids in Water at 25°C

Amino acid	pK_1	pK_2	pK_3	pI	Ref.
Alanine	2.35	9.87		6.11	3
Arginine	2.01	9.04	12.48	10.76	3
Asparagine	?	?	?	?	
Aspartic acid	2.1	3.86	9.82	2.98	3
Cysteine	1.71	8.33	10.78	5.02	5
Glutamic acid	2.1	4.07	9.47	3.08	3
Glutamine	2.17	9.13		5.65	5
Glycine	2.35	9.78		6.06	3
Histidine	1.77	6.10	9.18	7.64	3
Iso-leucine	2.32	9.76		6.04	3
Leucine	2.33	9.74		6.04	3
Lysine	2.18	8.95	10.53	9.47	3
Methionine	2.28	9.21		5.74	3
Phenylalanine	2.58	9.24		5.91	3
Proline	2.00	10.6		6.3	3
Serine	2.21	9.15		5.68	5
Threonine	2.63	10.43		6.53	5
Tryptophan	2.38	9.39		5.88	3
Tyrosine	2.20	9.11	10.07	5.63	3
Valine	2.29	9.72		6.00	3

Appendix F

The Structure of Common Alpha-Amino Acids

The general structure of the alpha-amino acids commonly found in proteins is as shown in Figure 9.24. With the exception of proline, all have a free carboxyl group and a free unsubstituted amino group on the alpha-carbon. They are distinguished by their different R groups. They may be classified according to the polarity of the R groups[5] into (1) nonpolar or hydrophobic R groups (Figure F.1), (2) uncharged polar R groups (Figure F.2), (3) R groups charged negatively at pH 6.0 (Figure F.3), and (4) R groups charged positively at pH 6.0 (Figure F.4).

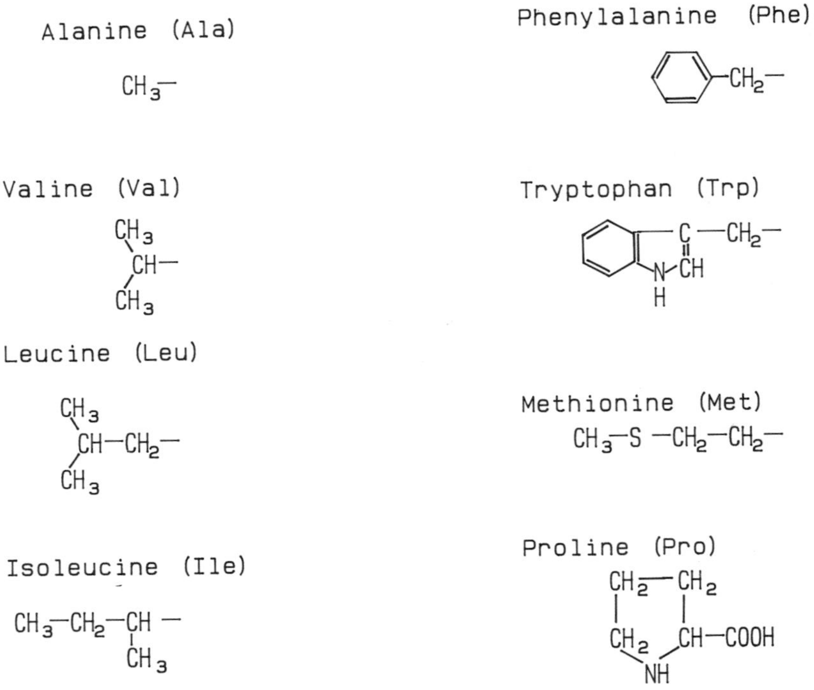
Alanine (Ala)
CH3—
Valine (Val)
CH3
CH—
CH3
Leucine (Leu)
CH3
CH—CH2—
CH3
Isoleucine (Ile)
CH3—CH2—CH —
CH3
Phenylalanine (Phe)
CH2—
Tryptophan (Trp)
C —CH2—
N-CH
H
Methionine (Met)
CH3—S —CH2—CH2—
Proline (Pro)
CH2—CH2
CH2 CH—COOH
NH

FIG. F.1. Amino acids and their R groups (nonpolar).

Glycine (Gly)

H —

Tyrosine (Tyr)

HO —⟨◯⟩— CH₂—

Serine (Ser)

HO —CH₂—

Asparagine (Asn)

$$NH_2 \diagdown \underset{O}{\overset{\diagup}{C}} -CH_2-$$

Threonine (Thr)

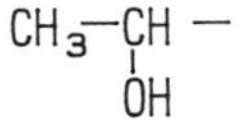

CH₃—CH —
      |
      OH

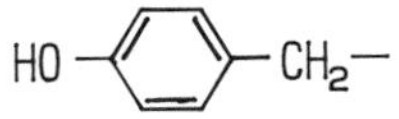

Glutamine (Gln)

$$NH_2 \diagdown \underset{O}{\overset{\diagup}{C}} -CH_2-CH_2-$$

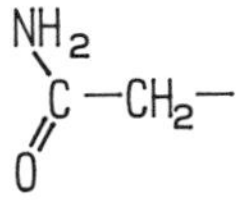

Cysteine (Cys)

SH —CH₂—

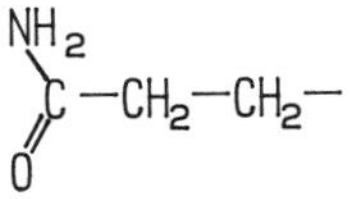

FIG. F.2. Amino acids and their R groups (uncharged polar).

Aspartic acid (Asp)

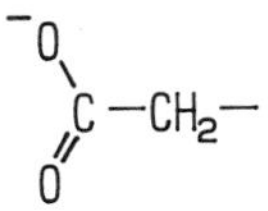

$$^-O \diagdown \underset{O}{\overset{\diagup}{C}} -CH_2-$$

Glutamic acid (Glu)

$$^-O \diagdown \underset{O}{\overset{\diagup}{C}} -CH_2-CH_2-$$

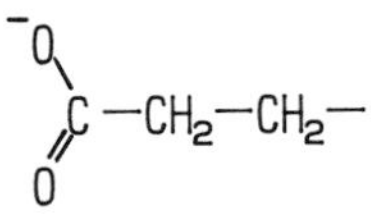

FIG. F.3. Amino acids and their R groups (negatively charged at pH 6.0).

Lysine (Lys)

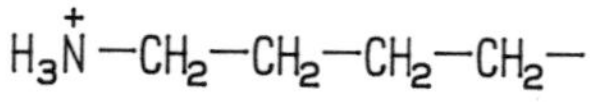

$$H_3\overset{+}{N}-CH_2-CH_2-CH_2-CH_2-$$

Arginine (Arg)

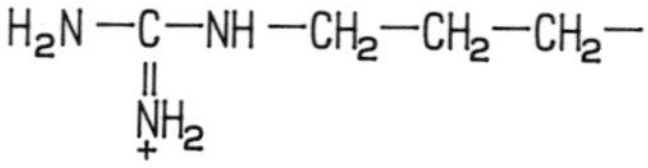

$$H_2N-\underset{\underset{+}{\overset{\|}{NH_2}}}{\overset{}{C}}-NH-CH_2-CH_2-CH_2-$$

Histidine (His)

$$\begin{array}{c} HC=\!\!\!=C-CH_2- \\ | \quad\quad | \\ H\overset{+}{N}\quad\; NH \\ \diagdown C \diagup \\ | \\ H \end{array}$$

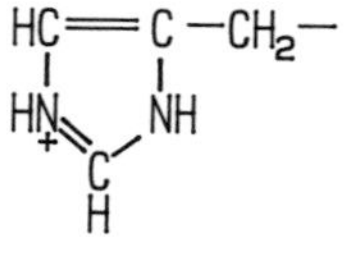

FIG. F.4. Amino acids and their R groups (positively charged at pH 6.0).

References for Appendixes

1. A. W. Adamson, *A Textbook of Physical Chemistry,* 2nd edition, Academic Press, New York (1979).
2. J. N. Butler, *Ionic Equilibrium. A Mathematical Approach,* Addison-Wesley, Reading, Massachusetts (1964).
3. *CRC Handbook of Chemistry and Physics,* 64th edition (R. C. Weast, ed.), CRC Press, Boca Raton, Florida (1983).
4. R. A. Robinson and R. H. Stokes, *Electrolyte Solutions,* Academic Press, New York (1955).
5. A. L. Lehninger, *Biochemistry,* 2nd edition, Worth Publishers, New York (1975).

Index